# COMBATING LONG-TERM UNEMPLOYMENT

# BRADFORD STUDIES IN EUROPEAN POLITICS

*Local Authorities and New Technologies*
Edited by Kenneth Dyson

*Political Parties and Coalitions in European Local Government*
Edited by Colin Mellors and Burt Pijnenburg

*Broadcasting and New Media Policies in Western Europe*
Kenneth Dyson and Peter Humphreys;
with Ralph Negrine and Jean-Paul Simon

# COMBATING LONG-TERM UNEMPLOYMENT

# LOCAL /E.C. RELATIONS

Edited by
Kenneth Dyson

**R**

**ROUTLEDGE**
London and New York

First published 1989
by Routledge
11 New Fetter Lane, London EC4P 4EE
29 West 35th Street, New York, NY 10001

© 1989 Kenneth Dyson

Printed and bound in Great Britain by
Mackays of Chatham PLC, Chatham, Kent

*British Library Cataloguing in Publication Data*

Combating long-term unemployment: local
   E.C. relations. — (Bradford studies in
   European politics)
   1. Long-term unemployment
   I. Dyson, Kenneth H.F. II. Series
   331.13′7

   ISBN 0-415-03127-3

*Library of Congress Cataloging in Publication Data*

Combating long term unemployment: local / EC relations / [edited by]
   Kenneth Dyson.
      p.  cm. — (Bradford studies in European politics)
    ISBN 0-415-03127-3
    1. Unemployment — European Economic Community countries.
   2. Manpower policy — European Economic Community countries.
   3. Labor policy — European Economic Community countries. I. Dyson,
   Kenneth H.F. II. Series.
   HD5764.5.A6064    1989
   331.13′77′094–dc19                      88–39349
                                                       CIP

# CONTENTS

# Contents

## FIGURES AND TABLES

### FIGURES

### TABLES

## CONTRIBUTORS

Stephen Barber, Head, European Social Fund Unit, Department of Employment

John Benington, Institute of Local Government Studies, University of Birmingham

Andrea Caspari, Local and Regional Development Planning, London

Kenneth Dyson, Professor of European Studies, University of Bradford

Derrick Johnstone, The Planning Exchange, Glasgow

Manfred Kaiser, Federal Institute for Employment, West Germany

David Kennedy, Head, Directorate for Employment and Environmental Services, Bradford

Haris Martinos, Local and Regional Development Planning, London

Derek Portwood, Unemployment Studies Unit, Wolverhampton Polytechnic

Tomas Roseingrave, Senior Visiting Research Fellow, University College, Dublin

*Contributors*

Jerry Sexton, The Economic and Social Research Institute, Dublin

Anna Whyatt, Chief Executive and Town Clerk, London Borough of Southwark

Ronald Young, Deputy Leader, Strathclyde Regional Council

x

# INTRODUCTION

Long-term unemployment is an intensely political subject, stirring deep human emotions especially on the part of those who have direct contact with the problem. Issues of social responsibility and human wastage, of social peace and care, are raised. This volume seeks to inform debate by analysing policy responses to the problem of long-term unemployment and by focusing on the role of local initiatives in a European context. Long-term unemployment is at once a global and European problem, needing general measures to alleviate it, and a group and local problem, being concentrated and unevenly distributed and needing intensive measures. It is at once an economic issue, involving questions of incentives and penalties, and a social issue involving marital and family stress and breakdown. Accordingly, policy responses need to be seen in their context, international and national, economic and ideological, and against the particular backgrounds and histories of local communities. Long-term unemployment is a major historical development affecting Western Europe in the final quarter of the twentieth century. Whole communities and families have already a history of long-term unemployment.

Long-term unemployment is one of the most intractable problems facing advanced industrialized societies. It reflects the greater inequality in their midst, between a 'richer rich' and a 'poorer poor'. On the one hand, there is a shrinking core of relatively privileged, highly professional, and spatially concentrated workers; on the other, an expanding social periphery comprising the old, the young, the unskilled, women, foreigners, and the handicapped. The existence of a large marginalized sector may act as a

1

powerful political incentive to maintain a measure of continuity in welfare-state provision (even when ideology points to radical change). In fact, there has been a restricted debate in Western Europe about radical change to welfare-state provision. Whilst the scope, volume, and timing of benefits and services are being reduced, programmes like those of unemployment benefit and basic health insurance are not being abandoned.

Denied the opportunities of both effective participation in the market and effective citizenship, and condemned to 'the curse of idleness', the long-term unemployed cannot remain a 'hidden' problem. The evidence of educational failure, broken families, and high crime rates will not go away; those who have no stake in 'official' society can make their presence felt in the form of spasmodic riot, delinquency, and violence. Even so, the real dimensions of the problem are easily lost when the majority basks in increasing material prosperity, with shorter working hours, longer holidays, and a perception of upward social mobility; and when so many of the long-term unemployed are concentrated in peripheral council-house estates.

In the 1980s the broad thrust of policy change towards the long-term unemployed has been provided by the 'conservative revolution' in economic ideology. The emphasis has been on tax cuts and incentives to cut unemployment and cutbacks in public expenditure and borrowing to promote investment and new jobs. In addition to control of inflation through monetary policy instruments, stress was placed on supply-side measures (like tax cuts, privatization, deregulation, and curbs on trade-union power) to improve the operation of the economy by creating an environment in which enterprise could flourish. Against this background trade unionism was in political and industrial retreat. Albeit on a small scale, industrial relations were refashioned with the spread of single-union deals and non-unionism. There was a drive for greater flexibility on payment systems, working methods, and ways of bargaining (notably with a trend to local bargaining and linkage of pay to local labour-market conditions). An overall fall took place in trade-union membership (in the UK down 20 per cent between 1979 and 1986, from 13.3 million to 10.5 million), whilst trade-union membership was strikingly lower in expanding areas like East Anglia, Greater London, and the South East, and in the expanding service sector. As a consequence, the prospects for trade-union influence on labour-market policies were reduced.

In fact, the role of the service sector as a job generator had already been seen in the United States where, between 1965 and 1986, service employment nearly doubled (38.8 million to 75.2 million). As a percentage of the non-agricultural total in the United States manufacturing employment had fallen from 34 per cent in 1950 to 19 per cent in 1986; service employment had increased from 59 per cent to 75 per cent. Labour-market conditions had not improved as a result. The service industries tended to pay low wages and were characterized by low productivity levels; the overwhelming majority of the new service jobs was in the low- or no-technology category (like fast food and retail); and the high-value-added jobs remained closely tied to manufacturing and tended to disappear or migrate with the latter. Comparative advantage in basic manufacturing was slipping to the newly industrialized countries, a transition in the world economy that was responsible for the phenomenon labelled in the 1970s 'de-industrialization'. This phenomenon was a key component of long-term unemployment.

The combination of rapid technological change with structural change within the economy exacerbated problems in the labour market. Micro-electronics and new communications technologies favoured capital-intensive rather than direct job-creating investment. In the process the labour requirements of manufacturing industry were reduced, with the expectation that the same process would characterize much of the service sector in the 1990s. Massive investment in training for skills and competence in the new technologies was essential if the potential, economic and human, of the new industrial revolution was to be realized. The threat of exclusion of a technologically illiterate population from participation in this revolution was very real. Again it surfaced in long-term unemployment.

Since the 1970s local authorities in Britain and other West European countries have been developing their own initiatives to combat some of the causes and consequences of long-term unemployment and associated poverty. This volume seeks, first, to encourage the pooling of experience of handling problems of long-term unemployment, particularly by learning from 'models of best practice' at the UK and European levels; and, second, to look at prospects, and problems, of European collaboration to combat long-

term unemployment and its relevance to local authority
strategy. Three specific themes are addressed:

> What can local authorities learn from experience of
> locally based initiatives elsewhere? How have such
> issues as developing new employment opportunities and
> integrating social provision (like housing, social
> services, and education) been tackled, and with what
> effects?

> Which developments at the EC level are most relevant
> to designing local strategies and measures to combat
> long-term unemployment? How is EC policy likely to
> develop in this field, and with what implications for
> local authorities?

> What has been the role of EC funding in the
> development of local initiatives in the UK? What kinds
> of initiative have emerged in the UK, and with what
> consequences?

This volume follows on from, and is a companion to,
Local Authorities and New Technologies: The European
Dimension. In its preparation I am indebted to the Goethe
Institute in Manchester, particularly its director, Gerhard
Murjahn, and to Wendy O'Conghaile of the European
Foundation for the Improvement of Living and Working
Conditions. Their help has been invaluable. The secretaries
in the Department of European Studies have once again
provided a superb service, particularly Jean Davison, Ele
Hay, Betty Jones, and Jennifer Kenyon.

The book draws together academics and practitioners
with an interest in European affairs and is based on a
conference held in January 1988 under the auspices of the
European Briefing Unit of the Department of European
Studies at the University of Bradford. It is designed to
appeal to all those interested in policies to combat long-
term unemployment - local- and central-government officials, private-sector organizations whose work is relevant to
locally based initiatives, and interested academics.

Kenneth Dyson

# Chapter One

# LONG-TERM UNEMPLOYMENT: THE IDEOLOGICAL PERSPECTIVE

Kenneth Dyson

From the mid-1960s onwards the body politic of Western Europe has experienced multiple shocks, leaving political actors, policy-makers, and professional analysts alike confused about their precise meaning and implications. Perhaps the most notable of these shocks are the following: the challenge to the post-war consensus on the managed economy and welfare state in the name of an 'anti-statist, anti-bureaucratic' politics associated with a new European intelligentsia (epitomized in the 'events of 1968'); the emerging political ascendancy of the new affluent middle classes which assumed a role-model character (embodied in the phenomenon of 'Thatcherism', neo-liberal political ideology, the new ruthless individualism and entrepreneurship of the 1980s, and the shift towards self-employment in the labour force); the threat of inflationary pressures to employment and competitiveness (represented by the two oil-price shocks of 1973 and 1978-9) and a consequent policy of budgetary rigour and public expenditure cuts; a mounting 'globalism' and rapidity of structural change across industrial sectors (reflected in 'deindustrialization' and industrial 'wastelands'); and an increase in poverty (the 1985 low-income figures for the UK showed 9.4 million people to be living on or below the 'poverty line', a rise of over a quarter since 1981). (1)

These phenomena were by no means unrelated to each other, and their economic, political, social, and cultural effects have been deep and complex. Nowhere have these effects been felt more deeply than in employment, the status of employment as an economic-policy aim and employment policies. On the one hand, the 1980s have seen an expansion of middle-class opportunities, a rising rate of social mobility and prosperity in small towns and rural areas. (2) On the other hand, long-term unemployment has become a key problem of the 1980s, at once pervasive and insidious and a symbol of the depth of the mounting social divide that besets Western industrial societies. In 1979 the long-term unemployed accounted for 25 per cent of all benefit claimants in the UK; by 1987 the proportion was 41 per cent. In economic policy full employment has either lost its pre-eminence as an aim or been seen as a residual effect of success in controlling inflation and appropriate 'supply-side' measures to deregulate and promote an 'enterprise culture'. With time the vision of full employment has lost credibility as a basis for policy. Reflecting wider conflicts about economic and social policy, employment policies have been characterized by a new dissensus. Just how high a priority should such policies enjoy, particularly in public expenditure? What form should they take: special employment measures (like the Community Programme), expansion of public services like education and health, targeted public investment (e.g. in construction) or 'workfare'? Sharp conflict seems the incontrovertible fact of life in employment policy in the 1980s. The question remains, nevertheless, of whether, from another perspective, the 1980s are a painful, slow, and disjointed learning process; of whether, beneath the conflict, there is a potential for a refashioned consensus in the 1990s around a new shared vision of the social-market economy. For the Right a shift towards such a consensus involves a recognition of the need for a 'human' face to capitalism, whether out of conscience or political expediency; for the Left it means acceptance of the market, enterprise, efficiency and private-public sector partnerships. By 1987-8 a measure of consensus had already emerged on the need to expand and improve official training and job schemes (e.g. a guaranteed place for every school leaver on the Youth Training Scheme in the UK). Will policy debate and conflict in the 1990s be taking place within a more radically altered but agreed basic framework - of a 'basic-income guarantee'; of a pragmatic acceptance of the need for a collaborative role of public and private sectors in providing 'high-quality' employment training and in matching skills to jobs, not least industry/education collaboration; and of a 'new pluralism' including an expanded role for local employment initiatives and for 'alternative' life styles and forms of production, and a new recognition of the idea of 'fulfilling' employment? As an ageing population places an impossible financial burden on health and social services, new value and incentive to community service is needed.

## THE NATURE OF THE PROBLEM

The long-term unemployed represent the enlarged number of victims of the processes of economic modernization and adjustment since the mid-1970s. Even the drop in the UK figure from April 1987 to April 1988 (from 1,295,146 to 1,029,206) did not reflect a greatly increased proportion of the existing long-term unemployed streaming out into employment. Two other factors were more significant: the application of stricter eligibility for work tests, thereby taking people off the unemployment register; and the greater number of the short-term unemployed finding work (so that they did not stream into the pool of long-term unemployment). At the heart of long-term unemployment (but by no means representing its full scale) is a lost generation of people who have been without work for over five years (271,000 in the UK in April 1988). Just as in technological and economic terms Western Europe seems to follow the United States, so - in social terms - the American phenomenon of an 'underclass' seems no longer a fanciful prospect for Western Europe. (3) All the evidence points to a widening of inequalities since the 1970s, with the 'safety net' for the poor in danger of falling into disrepair. (4) Social and communal values seem to be giving way to monetary rewards and motivations; the traditional Left's belief in the centrality of the one and indivisible working class has been radically challenged. (5)

Perhaps the most obvious consequence of this process of social division into 'insiders' (at work in the formal economy with secure jobs and growing earnings) and 'outsiders' (the unemployed and peripheral workers who live with insecurity and low incomes) are the problems of urban unrest, hooliganism, and violent disorder (e.g. Liverpool's Toxteth and London's Brixton), especially evident when unemployment is mixed with racial disadvantage and a sense of political powerlessness. (6) Less visible, but no less insidious, are the problems of the post-war housing estates in the urban peripheries and the disaster areas (in terms of jobs) to be found in the North, the Midlands and Scotland (notably Consett, Mexborough, and Hartlepool). By contrast, the 'job-generation honeypots' are to be found mostly in the South (Milton Keynes, Peterborough, Huntingdon, Basingstoke, Bracknell). (7) In other words, long-term unemployment is linked to the increasingly powerful image of a North-South divide in the 1980s. As some of the London boroughs and parts of Bristol reveal, this image can hide or gloss over important realities. (8) Even so, long-term unemployment within the UK is closely linked to the decay of distinctive regional economies, with tighter integration into the national and international economy, and with London as the 'core' and much else as the 'periphery'. (9)

The long-term unemployed are the victims of a restructuring of the labour market that has cut the availability of the jobs that they can perform. At the personal level the consequences are feelings of inadequacy, demoralization, increasing isolation, and loss of the confidence and will to keep looking for work; at the societal level there is the threat of a permanent underclass of disaffected, alienated onlookers, without work or income - without hope and lacking coping ability, money, stable domestic circumstances, literacy, and education. These unemployed people are denied the opportunities of both the market and effective citizenship, condemned to the curse of idleness, and contained by the twin weapons of minimum state welfare and efficient policing. (10) The profile of the long-term unemployed was well brought out in the UK Labour Force Survey of 1987: about one-quarter had literacy or numeracy problems; some 40 per cent had been in unskilled or semi-skilled manual jobs (compared with 22 per cent of the workforce in general); 49 per cent of adults seeking work for six months or more lived in council houses (compared with 19 per cent of economically active adults); especially affected were ethnic minorities with a poor knowledge of English, the disabled and the socially disadvantaged; and typically they lacked a car. Long-term unemployed people tended to be geographically immobile. Interregional migration of labour is mostly of the employed rather than the unemployed and of the young and the skilled; much of it is within the same company i.e. intrafirm mobility. (11)

The most severe problems of long-term unemployment were concentrated in the large metropolitan areas of Northern Britain hit by the sharp structural decline of a key industry like shipbuilding, steel, or coal. As a consequence, long-term unemployment had pronounced sectoral, regional, and local characteristics. Local and regional conditions, such as whether one lived in a larger town or city, the character of the local economic structure, and a family background of unemployment, could override the efforts of individuals to improve their situation on the basis of their personal attributes. Hence, for these reasons, effectively combating long-term unemployment meant focusing help locally. Because long-term unemployment was embedded in the nature of local culture and the strength of local institutions, a broad-based approach was required with a key role for local authorities as catalysts for cultural change, as training managers, and as integrators of social and economic support. As this volume reveals, many local authorities in the UK have been keen to seize and develop such a role. In the process they have been more discouraged than encouraged by British central government and more encouraged than discouraged by the European Commission. The consequence is an interesting process of forging new alliances between local and supranational actors, a process that contains within itself significant long-term political and cultural implications.

A profile of the long-term unemployed would be incomplete without reference to the motivation problem which arises from having been out of contact with the labour market for so long. This problem extends beyond simply the absence of relevant skills. Individual and group counselling is, accordingly, an indispensable part of the policy response. Self-confidence and awareness of one's rights must be generated if people are to be made available for work again; and, of course, rights (for instance, to child-care allowance for single-parent families) must be appropriate for this purpose. The unemployed must be treated as whole persons, taking great trouble over personal problems like child care. Key elements of an effective counselling system would include a programme of detailed individual interviews to tailor and structure help with job-finding and training (the purpose of the Manpower Services Commission's Restart Programme since July 1986); a sufficient number of claimant advisers (themselves with adequate computer backing on benefits); and special 'outreach' officers to contact people in difficult areas such as peripheral housing estates. In the United Kingdom nothing like sufficient resources have been devoted to this purpose; by 1988 the Restart Programme was suffering from a crisis of morale. Indeed efforts at counselling tended to be undermined by the pursuit of another purpose - getting people off the unemployment register by showing that they were not available for work. The consequence was to reinforce rather than resolve the motivation problem.

## THE NATURE OF THE POLICY RESPONSE

At the level of policy response an increasingly complex network of actors and complexity of programmes had arisen by 1988. Indeed the British government began to see complexity as a problem within its employment and training policies, bedevilling problems of delivery. A training industry had emerged, employing in the UK by 1987 some 23,000 people as managers, organizers, and other support staff: MSC, with 8,000 staff involved in employment, had roughly one member of staff for every 375 unemployed. Even so, the UK training industry remained very undeveloped compared with Sweden or West Germany, with private-sector involvement as a particularly weak link in the chain. The consequence is that apprenticeship schemes and craft training have been woefully neglected in the UK; over the last decade France, for instance, has forged ahead of the UK in the latter area. (12) Chambers of commerce (and some individual employers like Austin Rover) have woken up to the need for an increasingly active role for the private sector in training. They have. however, lacked the backing of an effective public policy for the purpose. It is scarcely surprising that Geoffrey Holland, as director of the MSC, argued: 'The real skills gap is that our country is under-educated, under-trained and under-skilled ... Our approach to training has for decades been too little, too narrow, for too few, with too many exceptions and with hardly any follow through.' UK employers have systematically undervalued training and been reluctant to assume responsibility for its improvement. (13)

Analysis of the development of policy in the UK must focus on the Manpower Services Commission (renamed the Training Commission in 1988). The MSC started life in 1973 to bring together the fringe activities of job centres and adult training. It grew in the late 1970s to become an unemployment-relief agency, spawning a series of special measures. By the late 1980s, and reflected in its change of name, the role was being redefined as a national economic agency to ensure the competence of all the workforce. In the process the MSC and its parent ministry, the Department of Employment, tried to broaden its role into policy areas previously dominated by local authorities and the Department of Education and Science. The MSC effectively assumed control over many vocational-training courses provided in colleges of further education. The £150 million Technical and Vocational Education Initiative extended its influence to the curriculum of secondary schools, with the aim of making the latter more work- and business-related. This growth went along with widespread scepticism about the role and effectiveness of the MSC and an uncertain political situation for two main reasons: local authorities and educationalists were often suspicious because the MSC has been used as a channel for funds outside the control of local councils and education authorities, thus displacing the scope for initiative through local politics; and, bringing together unions, employers, and government, the MSC was a corporatist body in an anti-corporatist age. It was also, in an age of expenditure cuts, an area of rapid growth; in 1986/7 the Conservative government spent some £1,700 million on training, over five times as much as was spent in 1978/9, the last year of the Labour government.

In addition to the pivotal role of the Training Commission and its bargaining relationship with the Department of Employment, the following actors are significant in the UK:

**The Department of Education and Science** has taken such initiatives as the development of a national curriculum with regular testing; the national Training and Vocational Education Initiative (TVEI) for secondary schools, launched in 1983 and intended to offer special courses for 14- to 18-year-olds that involved problem-solving skills and relevance to working life; the privately funded, more selective City Technology Colleges with a more technology- and business-oriented curriculum under the control of DES-appointed trusts and funded by private business; and, in adult education, Replan, a national programme to improve educational opportunities for unemployed adults. Local authorities lost control of the polytechnics with the 1988 Education Act. They became free-standing corporate institutions with boards again dominated by business interests, but were supervised by the DES. The idea of education/industry collaboration was further sponsored by groups like the Industrial Society and Business in the Community, particularly on the American Boston compact-scheme model dating from 1982. Such initiatives aim to ensure that skills (e.g. computing) within the local labour market match what local companies need; this is achieved by companies offering jobs in return for which they become involved in shaping the curriculum, setting standards, and improving the quality of schools. At the same time this process has undermined the autonomy of local education authorities who run the schools and (in England) non-university higher education.

**Higher- and further-education institutions** have played an increasingly important role, with further-education colleges responding in a new way to training needs and higher education as a catalyst for 'action research' with the unemployed. At the same time the expansion of special programmes like the Youth Training Scheme after 1983 gave the MSC a powerful lever over the further-education system.

**Local authorities** have a key role as training-management agents for the MSC and in developing strategies to combat long-term unemployment at the local level (see the chapters by Portwood, Kennedy, and Young). At the same time, heightened tensions and conflicts of interest were apparent between the MSC and many local authorities (e.g. Sheffield, Manchester, and some London boroughs), particularly about the Job Training Scheme (JTS) which was accused of devaluing training to 'mere work experience on the cheap'.

**Trade unions** had a significant role as 'partners' in the decision-making structure of the MSC and in helping to deliver training and work experience posts in the workplace. Hence government has continued to seek their approval for schemes; thus their non-co-operation with the JTS was a major blow for this scheme. At the same time efforts have been made to dislodge unions from the institutional positions which they occupied in the heyday of their power. For example, in 1982 16 statutory Industrial Training Boards were abolished and replaced by 100 non-statutory organizations. A system where the Trades Union Congress had a central role was replaced by one where the representatives of the Confederation of British Industry were dominant. The trade unions were divided over how far they should co-operate with the MSC's training schemes. The constraint on non-co-operation is that policy would be diverted to non-union workplaces exploiting cheap labour and the public sector's role in training provision would be diminished. Trade unions have also been active through the TUC Centres Against Unemployment.

Despite this range of actors the focus of action in policies towards the long-term unemployed remains the Training Commission. By 1988 MSC had experience of three main programmes in this area: the Community Programme, the Job Training Scheme, and Restart.

**The Community Programme** provided work for long-term unemployed people, full time or part time for up to a year, on projects that would benefit the community (e.g. gardening and decorating for old and disabled people, clearing up derelict land, tourism, and countryside projects, adapting buildings for community use, setting up creches and adventure playgrounds). By 1987 229,000 were employed on this programme. As the Community Programme was an employment initiative, training was not a requirement. Pay, though modest, was based on the local rate for the job.

**The New Job Training Scheme** had as its philosophy the idea of an individually designed package of structured training and linked practical experience with a local company to update existing skills or learn new ones. More controversial, especially with the trade unions, was the fact that it was the first example of a scheme in which the participants received only the same allowance as their benefit (plus travelling expenses). Principally because of trade union opposition the New JTS had only 24,000 participants by 1987 in place of the 100,000 anticipated. As a training scheme it seemed to perform badly. Of 7,175 trainees who left the scheme in January 1988, only 7.2 per cent had completed their training and only 17 per cent were known to have found work.

13

The Restart Programme offered, from April 1987, interviews at the local Jobcentre to everyone who had been out of work for six months or more. The aim was to help individuals take steps towards finding work, whether by a one-week Restart course to help with self-assessment, a place on the Community Programme, a place in a Jobclub (self-help groups for coaching in job-hunting techniques and support and motivation to search for work), the chance of self-employment with the Enterprise Allowance Scheme, a place on the New JTS or work on voluntary projects. By the end of 1987 only 0.5 per cent of interviewees under Restart had found employment.

## A COMPARATIVE PERSPECTIVE

It is helpful to put these initiatives of the MSC in a comparative framework to gain some idea of their adequacy. For this purpose, in line with the work of the Organization for Economic Co-operation and Development, five types of measure to assist the long-term unemployed can be distinguished. (14)

Placement and counselling activities The UK has notably lagged behind such countries as Sweden, the Netherlands, and France. In Sweden and the Netherlands a person who has been unemployed for six months is accorded priority in placement for job vacancies and on manpower programmes. In France, since 1982, the Agence Nationale Pour L'Emploi (ANPE) had conducted individual counselling on a regular and systematic basis (well before Restart in the UK); since 1985 modular training courses have been made available for this group.

Training programmes/retraining programmes At the forefront of development has been the well-established system of AMI-A centres in Sweden which mount special courses covering skill instruction, motivation, life skills, and job-search techniques; by 1983 20,000 participants were on courses lasting for an average of eleven weeks. The French modular training courses were taken by 70,000 in 1985 and lasted from 300 to 1,200 hours as required. Two distinctive innovations were the 'training workshops' or 'training firms'

14

pioneered by the state of North-Rhine Westphalia in 1978 and which have since spread across West Germany; and the 'training contracts' in France whereby firms are paid a subsidy if they undertake to train the long-term unemployed.

**Employment subsidies** This approach can involve cash payments to employers or reducing their wage costs (e.g. social-insurance contributions) and may or may not be targeted on specific groups (e.g. young people). Denmark has pioneered in the field of cash payments as part of the Job Offer Scheme; France has led the way in reducing social-insurance contributions. In this area the UK's Jobstart scheme of 1986 broke new ground by introducing an allowance paid directly to the individual who took a full-time job paying less than £80 per week. The aims were to encourage the long-term unemployed to make more intensive efforts to find work and to promote the acceptance of work involving lower rates of pay.

**Direct employment-creation programmes** These measures involve mainly public-sector agencies, increasingly working with voluntary and charitable bodies in either special employment projects or jobs created within existing public-sector agencies. In the field of special employment projects the UK's Community Programme has its equivalent in other EC states, typically sponsored by local authorities (nearly one-half of the work places on the Community Programme were provided by local authorities). Sweden and Belgium have been pioneering. By 1983 over one-third of all long-term unemployed people in Sweden participated in special employment schemes. Belgium introduced in 1982 the Troisième Circuit de Travail (Third Sector of Employment) to provide permanent jobs for the long-term unemployed, particularly in the area of personal and social services at the local level. Creation of jobs within existing public-sector agencies has been largely confined to programmes in the Netherlands and Denmark and again was focused on the local-authority level.

<u>Other activities</u> include:

**Programmes to encourage self-employment** Here, along with Denmark, the UK's Enterprise Allowance Scheme has been a pioneer in financial help to support unemployed people who want to become self-employed or start their own business (and have at least £1,000 available to invest in the business).

**Early retirement schemes** In 1984 France introduced 'solidarity allowances' for the long-term unemployed; the allowance payable (which is means tested) is doubled for beneficiaries over 55 years of age who have had 20 years of paid employment (or over 57.5 years if they have had 10 years of paid employment). These unemployed are effectively being placed in a retirement or pre-retirement category. More typical in Western Europe are examples like the UK's Job Release Scheme; financial help is given for older workers to retire early and release their full-time job to an unemployed person. Such vacancies are, however, unlikely to help the long-term unemployed.

**Programmes to encourage voluntary work** The Netherlands and the United Kingdom have pioneered the idea that the unemployed working on voluntary projects can continue to receive benefits (and in the case of the UK's Voluntary Projects Programme financial help was also given to the sponsors of the project). Projects include adult-education classes in social skills, literacy, numeracy, and computing; others involve local community activities such as environmental work or renovation; and some help people prepare for self-employment.

As one surveys the strengths and weaknesses in the UK's policy-responses to long-term unemployment, ideology appears as a powerful factor. The distinctive features of the UK's approach, from a comparative perspective, are the Enterprise Allowance Scheme and the Jobstart Allowance; less well developed in relation to the European 'leaders' are the placement and counselling services, training/retraining schemes, and special employment projects. The major overall UK weakness remains the training dimension where public-sector/private-sector collaboration is so vital and where the private sector has proved so reluctant to assume responsibility. At the same time this failure is not simply attributable to the ideology of the government. It is deeply rooted in a British political culture of arms-length relations between the public and private sectors. (15)

At the same time the emphasis on cross-national differences should not obscure common characteristics. Policies for the long-term unemployed in Western Europe have tended to be remedial, short term, and not sufficiently comprehensive in approach. They have been inappropriate to tackle a problem whose nature requires a multi-dimensional response, involving a combination of elements of intensive counselling and guidance, training, and work experience. Clear targeting has increasingly come to be seen as necessary. An aspect of this targeting is giving a stronger local dimension to policies for the long-term unemployed. (16)

## THE IDEOLOGICAL PERSPECTIVE OF THE NEW RIGHT

A distinctive feature of the 1980s has been the ideological renaissance of a neo-liberal view of the state, challenging the heritage of collectivist values associated with the traditions of conservative, christian-democratic and social-democratic thought in Western Europe. Prior to the 1970s variegated amalgams of these latter traditions had for long been ascendant in Western Europe. In a process that began to manifest itself in the 1970s and had even earlier origins, the traditions of conservative, christian-democratic and social-democratic thought were put on the defensive, indeed into retreat. This ideological shift has had enormous long-term implications for the character of policies towards unemployment, not least because neo-liberalism has powerful roots in changes in the social and economic structures of Western societies and in an international diffusion of ideas through transnational political and corporate structures. Emphasis shifted to the market as the guarantee of the efficient allocation of resources; to the individual as the ultimate repository of responsibility; to enterprise as the basis of wealth creation; and to wealth creation as the basis of the good society. (17) In attacking traditions in this way the ideologists of neo-liberalism became characterized as the 'New Right'.

The New Right argued that the welfare state was becoming an intolerable burden on the capitalist economy, manifested in growing disincentives to work and to invest. In their view the protracted economic recession, from the mid-1970s onwards, and the resulting mounting unemployment, were fundamentally caused by excessive state intervention. The welfare aims underlying the neo-Keynesian managed economy had created a high-cost, rigid industrial structure that was increasingly uncompetitive in international markets. Hence, confronted by the new challenge from Japan and the newly industrializing countries, Western economies were in structural crisis. They had, accordingly, radically to reconsider the relationship of state and society, including social policies.

The New Right offered an alternative to the 'bureaucratic state' and to 'corporatism' which were seen as the institutional expressions and supports of economic and social rigidities that hindered high productivity and efficiency. (18) In radically reducing the role of the state they were also creating the conditions for a stronger state. A target of New Right criticism was the 'over-extended' state which had taken on too many commitments, was prey to sectional pressures, and was losing authority. (19) The state must concentrate on fulfilling its basic functions: creating a framework of competition; maintaining order; and having a 'safety net' for the poor and disabled. Hence priority was accorded to deregulation and privatization in order to resuscitate market forces, to efficient policing, and to minimum state welfare provision.

In social policy the New Right focused on the transition from a 'benefit culture' (a culture of encouraging dependency by paying people to do nothing) to an 'enterprise culture' (in which the spirit of personal initiative and responsibility was encouraged and the principle of equality demoted). Emphasis was placed on the 'welfare burden' and the need to reduce the role of the state in welfare funding. Accordingly, public policy must narrow the definition of the needy and target support to this group better. The unemployed were to be encouraged to take part-time, low-paid work or places on special support measures like the Youth Training Scheme (YTS). Public policy had to move away from the 'gift' principle of Supplementary Benefit. The system of single grant payments for special needs was to be replaced by the repayable, discretionary 'loan' principle (as in the Social Fund in the UK, itself subject to cash limits and having as one of the tests of eligibility for loans 'the ability to repay'). Also, a greater role was to be restored to charitable organizations and the voluntary sector (organizations like the Family Welfare Association in the UK). This change could be achieved by using the 'carrot' (e.g. raising the limit on tax-allowable donations) or the 'stick' (e.g. the guideline for the UK's Social Fund that loans must not be given when another body has responsibility for providing the item or service in question). Above all, public policy must eliminate the 'unemployment trap' which occurs when benefits are set at a level that pays an individual to remain unemployed rather than seek work.

In relation to long-term unemployment New Right policies have taken three forms:

**The implementation of stricter availability-for-work tests** In the United Kingdom unemployment benefits are suspended if there are doubts about availability for work. Thus those not attending an invitation to a Restart interview for the long-term unemployed lose their benefit (approximately 10 per cent of cases); also, an attempt has been made to 'police' the register more effectively by tightening up the pre-interview questionnaire.

**'Workfare'** The idea of a 'work-for-benefit' system dates from the United States in the 1970s. Whilst 'workfare' has not become general policy in the United States and is not official policy in the United Kingdom, it has had considerable influence. The aim is to give people a push to get off welfare payments by providing incentives to work. Continuing payment of social-security benefits is tied to participation in a training scheme or taking up work that may be low paid (rather than tied to local or national wage rates). In the United Kingdom the £4 billion Employment Training Scheme, launched in September 1988, was seen as a move in the direction of workfare. In return for a place on this scheme, with an average of six months' training, participants would receive normal social benefit plus a top-up payment to reimburse for expenses (rather than 'the rate for the job' or a training allowance).

**The curtailed role of trade unions in vetting schemes** Trade-union power was seen by the New Right as the very heart of labour-market rigidities and consequent low productivity.

Correspondingly, the 1987 election manifesto of the British Conservative Party contained a commitment to increase the employers' influence over the MSC's decision-making machinery. A further step could be the withdrawal of government from most of its job-finding services in favour of private employment agencies.

In practice, even in a stronghold of New Right influence like the United Kingdom, these ideas proved difficult to implement. The idea of compulsory 'work for benefit' was both politically difficult to implement and dubiously practical. For these reasons even Boston's famous Employment and Training Choices (ET) programme was in fact voluntary. Workfare was ideologically problematic for libertarians, being too authoritarian for those attracted to neo-liberalism by its promise of personal freedom. (20) A neo-liberal could argue that in a civilized democracy people have the right to choose whether or not to work in the formal labour market and that, in choosing not to work in this sense, they maintain the unconditional right to a subsistence income. Workfare was also ideologically problematic for neo-liberals who emphasized law and order; increasing personal financial difficulties of the unemployed threatened an increase in crime, particularly youth crime. A further question involved the meaning and value of work in a work-for-benefit system. Workfare could mean cheap, forced labour (e.g. cleaning the streets, as in New York City's mandatory scheme). For a neo-liberal too it could be unacceptable, on grounds of human dignity and wider economic loss, for the unemployed to be forced into low-paid, menial jobs. In this sense workfare did not help the unemployed to improve their situation. On the other hand, workfare could mean sophisticated training schemes, supported by counselling and work experience.

Practical problems abounded. Pursuit of the implementation of stricter availability-for-work tests threatened to displace other goals, with serious internal repercussions in employment agencies as well as to the delivery of the overall service to the unemployed. Thus, in the United Kingdom Restart counsellors were placed under stress by contradictory pressures, especially by the mounting pressure to get people off the register. This situation militated against their counselling role; by 1988 Restart was finding work for less than 0.5 per cent of those interviewed. Resources were not available for detailed individual interviews, whilst there remained a lack of adequate job opportunities on offer. Adequate job and training opportunities raised in turn the practical questions of employers' response to workfare and of the costs of curtailing the role of the trade unions in employment policies. Quite simply, employers preferred to take on employees who wanted to do the job rather than those forced on them. Trade-union opposition could in turn reduce the value of work experience and training and thus undermine government policy at a time when increased priority was being given to overcoming skill shortages by greater emphasis on employment training. More fundamentally, interdependency remained a central characteristic of employment policies and counselled the pursuit of interactive decision-making rather than just confrontation. Such an approach was better adapted to the degree of fragmentation of power and to the effective mobilization of expertise in policy-making. The British government had also to draw the (costly) implications of taking new powers to disqualify people from unemployment benefit if not genuinely available for work - by guaranteeing a place on a special-measures scheme.

It is clear from the above analysis that ideological and practical problems have beset the implementation of New Right thought in the field of long-term unemployment. In practice continued extensive state provision of such 'public goods' as education, health services, and housing has been maintained. These problems and limitations suggest that such traditions as conservatism, social democracy, and christian democracy remain alive and influential. Equally, policy towards the long-term unemployed illustrates the success of the New Right in significantly shifting the boundaries of public debate and action. Its major impact has been on the terms and assumptions of policy debate.

## THE IDEOLOGICAL PERSPECTIVE OF THE NEW LEFT

The New Right has not been the only force for change at work in employment policies in Western Europe, as is apparent in the chapters by Roseingrave and Kaiser. Ideological change has also been under way within the West European Left. Two developments have underpinned this process: first, the collapse of belief in the industrial working class as the main agent of historical change, consequent on 'de-industrialization' and division into 'insiders' and 'outsiders', and leaving a void at the heart of the European Left; (21) and, second, the 'libertarianism of 1968', which unleashed a new critique of the state and of traditions of hierarchy and deference, and out of which emerged the 'alternative' movement (notably in West Germany and the Netherlands) and the theme of 'autogestion' (self-management) in France.

Against this background a new debate took shape about what to do with and for the poor and workless; about how far they need support, and how far they can assist themselves (see the chapter by Young for practical examples). Participation, self-expression, and 'fulfilling' employment began to replace the old faith of the Left in bureaucratic regulation, professional expertise, and monetary transfers. (22)

In the view of the New Left the traditional welfare state had itself contributed to present social crises. The welfare state was criticized for generating policy failure, political conflict, and social resistance. In the first place, it was trapped within an ideological contradiction: on the one hand, seeking to redistribute resources and opportunities; on the other, being constrained by the need to contribute to the efficient functioning of the market economy. (23) Second, the welfare state had actually generated new forms of social inequality and subordination by stigmatization of recipients, by inadequate and uneven provision, and by weakening clients' capacity for self-help in treating them as passive objects. Third, it had generated a conception of the state as an unlimited-liability insurance company, capable of underwriting all possible needs, risks, and failures. Fourth, the welfare state suffered from a chronic lack of co-ordination between various bureaucratic agencies, each safeguarding its own rationale and existence. A comprehensive view of the impact of the various activities of the state on an individual or group was not achieved in public policy. Muddle and 'ad hocery' prevailed. Fifth, the welfare state was overdependent for its effective operation on informal and unaccountable negotiations with powerful interest groups. Sixth, there was a new recognition that the structure of social and political support underpinning the traditional welfare state was vanishing. The working class was not a unitary and homogeneous force. Finally, against the backgrounds of technological changes (notably with micro-electronics) and of rapid structural change in the world economy, full employment in the traditional sense was losing its relevance as an inspiration for action.

The New Left's search for an alternative to the welfare state produced the idea of an egalitarian 'welfare society', decentralized in operation, and building on the new social movements (with their concerns about urban decay, environment, ecology, and minority rights) and 'post-material' values related to the quality of life. Social change was to be realized through a 'non-statist' strategy. The unemployed, the poor, and the low paid would themselves form the basis for new social initiatives such as producer, housing, and health co-operatives, neighbourhood organizations, and other grass-roots organizations (see for example the chapter by Young). In this move away from the vertically structured welfare state towards horizontally structured forms of mutual aid, local authorities had a special role in building up new networks of communication, advocacy, and mobilization. The New Left focused on the need to foster smaller-scale, co-operative social organizations. By encouraging direct and spontaneous modes of social action it would be possible to offset the depersonalizing effects of ever larger units and networks. The social basis of New Left strategy was provided not so much by the peripheral and underprivileged strata as by elements in the new middle classes, particularly in the professional and service sectors, with high levels of educational skills.

As far as long-term unemployment was concerned, the strategy of the New Left had two main implications: first, the importance of a new pluralism of institutional responses, with particular attention to local employment initiatives, alternative forms of production, and mutual aid and support (see for example the chapters by Young, Roseingrave, and Benington); and, second, the development of new forms of socially useful activity, not based on wage labour and the ideal of full employment, but on encouragement of a thriving 'informal' economy backed by a basic income guarantee.

This redefinition of the terms in which productivity and utility are defined found during the 1970s and 1980s a new objective basis in the depletion of raw materials and environmental despoilation and in the impact of new technologies on employment. New Left thought linked up also with powerful libertarian traditions in Western Europe.

## CONCLUSION

By the end of the 1980s New Right thinking, and to an extent New Left thinking, had had a significant impact on policy debate about long-term unemployment, in the United Kingdom and in the major industrial countries of the European Community. In so far as they had brought about a revolution, it was partial - in part because of the tenacity of conservative, social-democratic and christian-democratic traditions, and in part because of the practical problems of implementation that they themselves engendered and the tenacity and influence of interests and rules in policy-making. The fundamental change was in the climate within which policy had to operate. Consensus seemed the chief victim. The overall impression was of ideological conflict, programmatic complexity, and political and administrative 'ad hocery'. Policy initiatives appeared as a series of 'leaps in the dark', stronger on political gesture than careful research into who wins and loses in the process.

At the same time it can be argued that this situation masks a slow, complex, and tortuous process of policy learning and development. The realities of fragmentation and interdependency in policy-making and particularly implementation for the labour market intrude: interdependency of centre and locality, of public and private sectors, and of both sides of industry. Perhaps more importantly, a critique of the traditional welfare state is associated with both New Right and New Left. When a fundamental debate gets under way about the future structure of the welfare state, the idea of a basic minimum-income guarantee provides an opportunity to forge a new consensus. This theme will be returned to in the concluding chapter.

Nevertheless, similarities in the acceptance of national responsibility for the long-term unemployed must not obscure the role and impact of national intellectual traditions and institutions. Individual EC states have moved at different speeds and in different ways. (24) Thus West German policies are integrated more solidly into the political and social systems than Britain's. They are, as a consequence, less isolated and controversial. Whereas British policies have been largely shaped by a preoccupation with poverty and, increasingly, efficiency, West German policies have always been seen as part of the wider problem of how best to bring about political integration - hence the preoccupation with the terms 'solidarity' and 'social peace' - and shaped by bargaining between different levels of government and different parties. Not only Scandinavian policies but also the French approach to long-term unemployment gravitate closer to West Germany's than to Britain's approach. Increased opportunities for the diffusion of experience, in particular through the EC, must not obscure the fine details of political and socio-economic values and forces and the way in which these values and forces inform the texture of programmes, the fine print of rules and regulations.

## REFERENCES

1    On the New Left see C. Offe (1984) Contradictions of the Welfare State, London: Hutchinson, and G. Hodgson (1984) The Democratic Economy, Harmondsworth: Penguin. On the New Right see D. Green (1987) The New Right, Brighton: Wheatsheaf; R. Plant and K. Hoover (1988) The Rise of Conservative Capitalism, London: Methuen; S. Brittan (1983) The Role and Limits of Government, London: Temple Smith; N. Bosanquet (1983) After the New Right, London: Heinemann; D. King (1987) The New Right: Politics, Markets and Citizenship, London: Macmillan, and N. Barry (1987) The New Right, London: Croom Helm. On economic policy see (1988) Economic Policy: The Conservative Revolution, Cambridge: Cambridge University Press, and A. Cox (ed.) (1984) Politics, Policy and Economic Recession, London: Macmillan. On structural change see M. Beenstock (1982) The World Economy in Transition, London: Allen and Unwin, and K. Dyson and S. Wilks (eds.) (1983) Industrial Crisis, Oxford: Martin Robertson. On poverty see N. Deakin (1987) The Politics of Welfare, London: Methuen; J. Mack and S. Lansley (1985) Poor Britain, London: Allen and Unwin, and Social Trends 17 (1987), London: Central Statistical Office, HMSO.
2    See Social Trends, op. cit.
3    K. Auletta (1982) The Underclass, New York: Random House and J. Rentow (1987) The Rich Get Richer, London: Allen and Unwin.
4    J. Rentow, op. cit.
5    C. Offe, op. cit.; A. Gorz (1982) Farewell to the

Working Class, London: Pluto.

6   J. Benyon and S. Solomos (eds.) (1987) The Roots of Urban Unrest, Oxford: Pergamon.

7   J. Goddard and M. Coombes (1987) The North-South Divide: Local Perspectives¨ Newcastle: Centre for Urban and Regional Development Studies, University of Newcastle.

8   N. Buck and I. Gordon (1986) The London Employment Problem, Oxford: Clarendon Press.

9   R. Martin and R. Rowthorn (eds.) (1986) The Geography of Deindustrialization, London: Macmillan.

10  S. McRae (1987) Young and Jobless: The Social and Economic Consequences of Long-Term Youth Unemployment, London: PSI. Also D. Ashton and M. Maguire (1988) Young Adults in the Labour Market, London: Department of Employment.

11  A. Champion, A.E. Green, D.W. Owen, D.J. Ellis, and M.J. Coombs (1987) Changing Places: Britain's Demographic, Economic and Social Complexion, London: Edward Arnold.

12  See various reports in the National Institute Economic Review, especially the work of S. Prais. Also A. Ovenden (1987) Competitiveness in UK Manufacturing Industry, London: BIM, and G. Worswick (ed.) (1986) Education and Economic Performance, Aldershot: Gower.

13  Competence and Competition, London: NEDO Books. For the Holland quote see C. Leadbeater, (1987) 'When it's time to stop passing the buck', Financial Times 7 September.

14  Organization for Economic Co-operation and Development (OECD) (1986) Employment Creation Policies: New Roles for Cities and Towns, Paris: OECD and especially OECD Manpower and Social Affairs Committee (1987) Measures to Assist the Long-Term Unemployed, Paris: OECD.

15  K. Dyson, 'The cultural, ideological and structural context', in K. Dyson and S. Wilks (eds.), op. cit.

16  J. Chandler and P. Lawless (1985) Local Authorities and the Creation of Employment, Aldershot: Gower.

17  See note 1. above. Also P. Cosgrove (1978) Margaret Thatcher, London: Heinemann; Sir K. Joseph (1975) Reversing the Trend, London: Centre for Policy Studies; M. Thatcher (1977) Let Our Children Grow Tall; London: Centre for Policy Studies; and P. Minford

(1984) 'State expenditure: a study in waste', Economic Affairs, vol. 4, no. 3. On a philosophical plane see especially R. Nozick (1971) Anarchy, State and Utopia, New York: Basic Books.

18  M. Olson (1982) The Rise and Decline of Nations, New Haven: Yale University Press.

19  S. Brittan (1977) The Economic Consequences of Democracy, London: Temple Smith, and A. King (ed.) (1976) Why is Britain Becoming Harder to Govern?, London: BBC.

20  S. Brittan (1987) 'Capitalism and the under class', Financial Times, 1 October.

21  E.g. C. Offe, op. cit.

22  See e.g. C. Offe (1986) Disorganized Capitalism, London: Polity Press; and J. Keane and J. Owens (1986) After Full Employment, London: Hutchinson.

23  J. O'Connor (1973) The Fiscal Crisis of the State, London: Macmillan.

24  D. Ashford and E. Kelley (eds.) (1986) Nationalizing Social Security in Europe and America, London: Jai Press. On France in particular and a useful Anglo-French comparison see D. Ashford (1986) The Emergence of the Welfare States, Oxford: Blackwell.

**Chapter Two**

**LONG-TERM UNEMPLOYMENT: THE INTERNATIONAL PERSPECTIVE (1)**

Jerry Sexton

Before I discuss the long-term unemployment problem in any detail, it is first of all appropriate to consider the question as to how it should be defined. Currently, if one takes note of recent government reports, the research literature on the subject, and such aspects as the eligibility criteria which have to be met for participation in state manpower schemes, a minimum duration of one year's unemployment would appear to be the generally accepted convention. This is, in fact, the definition which is adopted throughout this chapter.

It should be mentioned, however, that the concept of what constitutes long-term unemployment has tended to change over time. In his comprehensive 1968 report on long-term unemployment in OECD countries, Sinfield used a definition based on a minimum duration of six months. However, the increasing severity of the employment problem in many countries has created a climate where relative views have changed. Six months' unemployment experience is now no longer considered to be as serious a matter as it once was, when viewed against the background of the sizeable numbers who have been without work for much longer periods. The long-term unemployment position has continued to worsen in terms of average duration and, as time progresses, one is motivated to conceive or conceptualize long-term unemployment in terms of increasing time horizons.

*The International Perspective*

## THE EXTENT OF LONG-TERM UNEMPLOYMENT IN DIFFERENT COUNTRIES

The position regarding the extent of long-term unemployment varies considerably between countries. This can be seen from table 2.1 which shows relative data for 1985 on both unemployment and long-term unemployment for selected OECD countries. The table includes all the countries of the European Community and these are also grouped together in order to facilitate comparisons between countries within the Community, and between the Community and other important Western economies.

Observing the data on long-term unemployment (which are expressed in the form of the numbers of long-term unemployed taken as a proportion of total unemployment) within the European Community, one can identify three broad Community groupings. In the first instance, there are countries like Denmark and Luxemburg with a relatively low degree of long-term unemployment, the proportions of total unemployment being less than 40 per cent. Then there is a larger group comprising Germany, France, the United Kingdom, and Greece with long-term unemployment shares of between 40 and 50 per cent, while finally Ireland, Belgium, the Netherlands, Italy, Spain, and Portugal all exhibit relatively high shares of long-term unemployment, the proportion being well in excess of 50 per cent in each case. Belgium has the highest proportion with a percentage of just less than 70.

The aggregate proportion of long-term unemployment in 1985 for the twelve countries which now form the European Community was 54 per cent. This, it should be noted, relates to an absolute Community total of some 7.4 million persons without work for at least a year, compared with a total unemployment level of 13.8 million relating to those for whom information on duration of unemployment was available or relevant. (2)

The above-mentioned relative long-term unemployment figures are much higher than those indicated for the other OECD countries listed in the table, which basically relate to North America, Scandinavia, as well as Australia. Among these countries Australia exhibits the highest incidence of long-term unemployment but even this, at 31 per cent, is considerably less than the lowest Community country figures of some 38 per cent (for Denmark and Luxemburg). For the other non-EEC countries listed, the proportions in

29

Combating Long-term Unemployment

**Table 2.1** Unemployment and long-term unemployment in selected OECD countries in 1985

| Country | Proportion of long-term unemployed in total unemployment | Unemployment rate |
|---|---|---|
| | % | % |
| Belgium | 69.2 | 11.3 |
| Denmark | 38.5 | 7.8 |
| FR Germany | 47.5 | 6.9 |
| Greece | 46.3 | 7.8 |
| France | 46.8 | 10.3 |
| Ireland | 64.1 | 18.0 |
| Italy | 65.8 | 9.2 |
| Luxemburg | (37.1) | 3.0 |
| Netherlands | 58.7 | 10.5 |
| United Kingdom | 48.1 | 11.5 |
| Europe of the 10 | 52.9 | 9.5 |
| Spain | 57.8 | 21.9 |
| Portugal | 56.4 | 8.6 |
| Europe of the 12 | 54.0 | 10.7 |
| Australia | 30.9 | 8.2 |
| Canada | 10.3 | 10.4 |
| United States | 9.5 | 7.1 |
| Sweden | 11.4 | 2.8 |
| Norway | 8.3 | 2.5 |

Sources etc.: All the long-term unemployment ratios have been derived from Labour Force Surveys. Those for EEC countries (with the exception of Spain and Portugal) have been taken from the 1985 EUROSTAT Community Labour Force Survey. Those for other countries were obtained from the annual OECD Employment Outlook series.

The unemployment rates have been taken from the OECD Employment Outlook publication for 1987. With the exception of the rates for Denmark, Greece, Ireland Luxemburg, and Portugal these are standardized rates which have been adjusted in accordance with the methods outlined in the 1985 OECD publication Standardized Unemploymen Rates - Sources and Methods.

question all lie in the 5 to 15 per cent range.

## RECENT TRENDS IN LONG-TERM UNEMPLOYMENT

All countries, irrespective of their current position regarding the degree or level of long-term unemployment, have experienced an escalation in the problem over the last decade. This is illustrated in table 2.2 which gives country figures for both the proportion of long-term unemployment and unemployment rates for the period from 1975 to 1986.

One can subdivide the growth in long-term unemployment into two stages. The first followed the 1974-6 recession which, broadly speaking, resulted in a doubling of the incidence of long-term unemployment in most countries. The second escalation occurred subsequent to the renewed onset of recession in 1979/80 following the second oil-price shock. Initially this gave rise to the emergence of sizeable numbers of new unemployed which, for a while, caused the measured proportion of long-term unemployment to fall - as the figures for 1981 in table 2.2 illustrate. However, as the latter downturn deepened and persisted, the numbers out of work for long periods grew and the share of long-term unemployment in the total unemployment stock resumed its upward climb. The more recent figures for 1986 and the immediately preceding years are of particular interest since, by this time in a number of countries, the unemployment rates had begun to stabilize or fall. It is noticeable, however, that in many countries, and in Community countries in particular, this overall improvement did not carry over into any significant alleviation of long-term unemployment. Denmark provides the most notable European example in this context. The Danish unemployment rate began to fall significantly from 1983 onwards, but the proportion of long-term unemployment continued to rise steadily. The same phenomenon occurred in Canada. However, in some other countries (United States, Sweden, Norway) the fall in the unemployment rate was followed by a decline in the proportion of long-term unemployment, even if the latter occurred with some delay. It will be of interest, therefore, to observe whether the trend of long-term unemployment in the European Community will show any improvement over the next year or so. In effect, one needs to wait until an economic cycle has worked itself through and the dust has

**Table 2.2** Unemployment rates and long-term unemployment as a proportion of total unemployment in Community countries and in selected OECD countries over the period 1975 to 1986

| Country | Unemployment rate | | | | | | | |
|---|---|---|---|---|---|---|---|---|
| | 1975 % | 1979 % | 1981 % | 1983 % | 1984 % | 1985 % | 1986 % |
| EC countries | | | | | | | |
| Belgium | 5.0 | 8.2 | 10.8 | 12.1 | 12.1 | 11.2 | 10.8 |
| Denmark | 4.9 | 6.0 | 10.3 | 11.4 | 8.5 | 7.3 | 6.3 |
| FR Germany | 3.6 | 3.2 | 4.4 | 8.0 | 7.1 | 7.2 | 6.9 |
| Greece | 2.3 | 1.9 | 4.1 | 7.9 | 8.1 | 7.8 | 7.4 |
| France | 4.6 | 5.9 | 7.3 | 8.3 | 9.7 | 10.1 | 10.3 |
| Ireland | 7.3 | 7.1 | 8.9 | 14.0 | 15.6 | 17.4 | 18.0 |
| Italy | 6.2 | 7.6 | 8.3 | 9.8 | 10.2 | 10.5 | – |
| Luxemburg | 0.2 | 0.7 | – | 1.6 | 1.7 | 1.6 | 1.4 |
| Netherlands | 5.2 | 5.4 | 8.5 | 12.0 | 11.8 | 10.6 | 9.9 |
| United Kingdom | 4.3 | 5.0 | 9.8 | 12.5 | 11.7 | 11.2 | 11.1 |
| Spain | 3.7 | 8.5 | 14.0 | 17.2 | 20.1 | 21.4 | 21.0 |
| Portugal | 4.4 | 8.0 | 8.1 | 8.2 | 8.9 | 9.0 | 8.8 |

Table 2.2 continued

|  | Unemployment rate | | | | | | |
|---|---|---|---|---|---|---|---|
| Country | 1975 | 1979 | 1981 | 1983 | 1984 | 1985 | 1986 |
|  | % | % | % | % | % | % | % |
| Other countries |  |  |  |  |  |  |  |
| Australia | 4.8 | 6.2 | 5.7 | 9.9 | 8.9 | 8.2 | 8.0 |
| Canada | 6.9 | 7.4 | 7.5 | 11.8 | 11.2 | 10.4 | 9.5 |
| United States | 8.3 | 5.8 | 7.5 | 9.5 | 7.4 | 7.1 | 6.9 |
| Sweden | 1.6 | 2.1 | 2.5 | 3.5 | 3.1 | 2.8 | 2.7 |
| Norway | 2.3 | 2.0 | 2.0 | 3.3 | 3.0 | 2.5 | 1.9 |

Table 2.2 continued

| | Long-term unemployment as proportion of total unemployment | | | | | | |
|---|---|---|---|---|---|---|---|
| | 1975 | 1979 | 1981 | 1983 | 1984 | 1985 | 1986 |
| EC countries | % | % | % | % | % | % | % |
| Belgium | 29.7 | 61.6 | - | 65.8 | 68.4 | 69.2 | 70.3 |
| Denmark | 9.4 | 36.2 | 26.2 | 32.2 | 36.9 | 38.5 | - |
| FR Germany | 11.8 | 28.7 | 22.4 | 39.3 | 45.1 | 47.9 | - |
| Greece | - | - | 21.9 | 34.8 | 39.1 | 46.2 | - |
| France | 16.3 | 31.0 | 33.4 | 42.5 | 42.1 | 46.8 | 47.8 |
| Ireland | 19.1 | 38.2 | - | 36.4 | 45.8 | 63.8 | 64.5 |
| Italy | 33.8 | 51.2 | 50.3 | 56.3 | 62.1 | 65.7 | - |
| Luxemburg | - | (22.8) | - | (33.9) | (27.4) | (37.5) | - |
| Netherlands | 18.6 | 35.9 | 29.6 | 49.4 | - | 58.7 | - |
| United Kingdom | 14.8 | 29.5 | 29.3 | 46.6 | 48.1 | 48.1 | 45.0 |
| Spain | - | 27.5 | 43.6 | 53.6 | 54.2 | 57.8 | 56.6 |
| Portugal | - | - | - | - | - | 56.4 | - |

Table 2.2 continued

| Other countries | Long-term unemployment as proportion of total unemployment | | | | | | | |
|---|---|---|---|---|---|---|---|---|
| | 1975 | 1979 | 1981 | 1983 | 1984 | 1985 | 1986 | |
| | % | % | % | % | % | % | % | |
| Australia | - | 18.1 | 21.1 | 27.5 | 31.2 | 30.9 | 27.5 | |
| Canada | 1.3 | 3.5 | 4.2 | 4.8 | 10.1 | 10.3 | 10.9 | |
| United States | 5.3 | 4.2 | 6.7 | 13.3 | 12.3 | 9.5 | 8.7 | |
| Sweden | 6.2 | 6.8 | 6.0 | 10.3 | 12.4 | 11.4 | 8.0 | |
| Norway | - | 3.8 | 3.0 | 6.7 | 10.8 | 8.3 | 6.7 | |

Sources, notes: The unemployment rates have been taken from various issues of the annual OECD Employment Outlook publication; with the exception of the rates for Denmark, Greece, Ireland, Luxemburg, and Portugal. These are standardized rates as explained in the notes to table 2.1

The long-term unemployment ratios, which are all based on Labour Force Survey sources have been taken from the EUROSTAT Annual Labour Force Survey reports, the OECD Employment Outlook publications and the reports of national Labour Force Surveys.

There are some discontinuities in the figures which should therefore be taken as only indicating trends of a broad nature.

settled in order to assess the full extent of the human damage caused in the form of long-term unemployment. At present there are few positive signs, except in the United Kingdom where the proportion of long-term unemployment in total employment began to decline in 1986.

It is necessary to emphasize that, in the context of current developments in society, one does not necessarily obtain a complete picture of the changing labour-market scene by relying solely on conventional measures of unemployment. In addition to those classified as unemployed there are now millions throughout Europe accommodated on manpower schemes and many more have been pushed beyond the peripheries of the core labour market (e.g. those classed as 'discouraged' or 'underemployed'). In a wider social context one really needs to have a knowledge of how these additional groups have expanded or contracted if a comprehensive view of the full impact of the employment crisis is to be obtained. One must thus now view any apparent improvements in labour conditions as indicated by conventional unemployment measures with a degree of circumspection.

## CAUSATIVE FACTORS

Basically a major factor contributing to both unemployment and long-term unemployment alike in many countries has been an insufficient level of demand arising from both sluggish economic growth and the changing patterns in the job market. No one feature has of itself, however, created the problem, and one must look to a combination of factors in order to postulate reasonable explanations for its growth in recent years. The fact that in many countries the employment crisis has persisted for so long in a period of continuing labour-force expansion is an aspect that should be particularly noted. Increasing numbers of entrants to the labour market accentuate the problems of oversupply. As a result the existing unemployed, particularly those who are less well equipped and unsuited to the vacancies on offer, get pushed further down the queue of jobless and gradually drift into long-term unemployment, or perhaps out of the labour force altogether as they become discouraged and their propensity to seek work declines.

This 'queuing' phenomenon tends to reinforce further the long-term unemployment problem since, once a person becomes associated with the stigma of long-term unemployment, it is even more difficult to escape from that state. This aspect (referred to as the 'duration dependence' factor) is in fact one of the most widely verified features of long-term unemployment. Thus long-duration unemployment involves a kind of entrapment, since in filling vacancies employers tend to select from the more recently unemployed, better motivated applicants, thus leaving the long-term unemployed (particularly those in the older age groups) in an increasingly disadvantageous position.

A further factor which has contributed to long-term unemployment is the changing pattern of employment demand, both in terms of sectoral shifts and the different types of skill which characterize the areas where employment expansion is now concentrated. Sectoral or occupational restructuring is, of course, an ongoing process and has been evident in Western economies for some time, particularly in regard to the decline in employment in industry and the rise in the numbers at work in the tertiary sector. Throughout all sectors there has been a decline in the demand for unskilled labour. In recent years, however, in the wake of the recession which followed the second oil-price shock of 1979, the process appears to have accelerated. The result has been large-scale redundancies in traditional industries such as clothing, textiles, coal mining, heavy engineering, shipbuilding, etc. This scenario, when viewed in the context of other factors such as the 'queuing' phenomenon just referred to, is obviously one which can lead to a serious long-term unemployment problem. In such circumstances redundant workers find it difficult to avail themselves of new job opportunities which are frequently in different locations and require higher educational levels and/or different skills and attitudes.

The wide variation across countries, evident from the statistical indicators given in table 2.1, suggests that basic differences in the way labour markets operate can influence the unemployment situation. In this context labour-market flexibility is an issue which inevitably comes to mind. The view is held that lower levels of both unemployment and long-term unemployment are more easily attained in more fluid, less regulated labour markets. Opinions differ, however, as to the significance of this argument. Proponents of the flexibility argument point to the experience in North American labour markets and attribute the low unemployment levels achieved to the more fluid conditions which prevail, characterized by greater mobility and higher flows into and out of both employment and unemployment. This is contrasted with the consistently higher unemployment rates and higher incidence of long-term unemployment in many European countries. There are, of course, exceptions to this general pattern. The figures presented earlier indicate that unemployment is relatively low in the Swedish and Norwegian labour markets, both of which, it should be noted, involve a significant degree of regulation.

However, the very concept of flexibility in the labour market is a highly complex area embracing many strands such as labour mobility, job security provisions, labour cost flexibility (both in aggregate and sectorally), the relationship between wages and unemployment benefits, adaptability within enterprises, etc. The only existing study which attempts to deal comprehensively with a range of such aspects is a recent OECD Report (1986) on this issue. The conclusions of this report, while confirming that there is a significant relationship between important aspects of flexibility and variations in the level of employment, also recognize that many other factors (such as changes in the level of aggregate demand) impinge on the situation. The report thus stops short of concluding that flexibility is a major explanatory factor underlying the differences in the degree and pattern of unemployment across countries.

## MEASURES TO DEAL WITH LONG-TERM UNEMPLOYMENT

I now wish to consider briefly the question of the measures used to counter long-term unemployment. The various activities which I will discuss are not, of course, all designed solely with a view to alleviating long-term unemployment. Broadly speaking, they are general-purpose measures designed to assist all unemployed persons, even though some of them involve targeting arrangements aimed at specific groups, such as the long-term unemployed. The programmes in question can be broadly categorized under four headings:

Special placement and counselling activities.
Measures designed to achieve the direct integration of the long-term unemployed into employment.
Training programmes.

Direct employment-creation measures.

There are, of course, other kinds of programme. However, it can generally be said that measures of the above types account for the great majority of persons accommodated on manpower schemes.

Before, however, I engage in any discussion or summary evaluation of these activities, it is perhaps first of all appropriate to review briefly the background to these measures, and the context in which they have evolved over the last two decades. This is necessary as, in initiating new approaches or strategies, one is rarely in a position to start with a clean slate; inevitably the status quo imposes some constraints in regard to the possible options.

Generally speaking, the measures covered by the above-mentioned headings grew out of the concept of an active manpower policy as articulated in the OECD recommendations of 1964. These involved an emphasis on labour supply-side policies relating to training, placement, and mobility measures in order to meet the needs of expanding labour demand, and the parallel restructuring which accompanied this growth. This type of strategy was appropriate in the buoyant conditions of the 1960s and early 1970s. However, the scene changed drastically in the mid-1970s with the onset of the severe 1974/6 recession which gave rise to then unprecedented levels of unemployment and, simultaneously, high inflation. A growing view that manpower-supply measures were of reduced effectiveness during a prolonged period of sluggish economic growth and slack labour-market conditions led to a change in emphasis. Thus the concept of a 'General Employment and Manpower Policy' which was to be integrated into the overall framework of economic and social policies was devised and was embodied in further recommendations formulated by the OECD in 1976. In these new guidelines a higher profile was given to the use of selective employment and manpower measures with the objective of achieving and maintaining high levels of employment, but in ways which did not contribute to creating inflationary pressures. As a result emphasis was placed on measures such as marginal employment subsidies. The equity aspect was also accorded greater importance in the 1976 recommendations in so far as countries were recommended to render specific assistance to 'disadvantaged groups to enter, remain or return to employment, thereby promoting more equity in the distribution of employment opportunities and income'.

It should be borne in mind, however, that the 1976 OECD recommendations reflected a shift in emphasis rather than a fundamental change. The rationale which underpinned manpower measures across many countries remained basically the same. This applied to objectives and administrative provisions and, understandably, influenced the perceptions of those who operated the schemes. As such, the general body of manpower programmes which existed in most countries was particularly unsuited to dealing with the large numbers of disadvantaged unemployed which emerged in the aftermath of the renewed onset of recession in 1979/80. This explains why many of the initiatives subsequently taken, even though substantial in many countries, have tended to be of an ad hoc nature and have not been carried out in the context of an overall framework which gives due recognition to the fundamental changes which have occurred in the labour market. One of the few exceptions here relates to Denmark where a comprehensive programme to provide temporary employment for the long-term unemployed was introduced as far back as 1978 in the form of the Danish Job Offer Scheme.

When it became clear that long-term unemployment was emerging as a serious problem, in a number of countries the first approach to countering the problems involved more intensive counselling and placement activities. France, for example, mounted a major initiative of this kind in 1982. 'Operation Long-Term Unemployed' can only be described as a massive blitz on the problem; it involved interviewing and assessing every one of the 350,000 long-term unemployed persons on the unemployment register between October 1982 and March 1983. The ultimate aim was to gain additional insights into the nature of the problem and to facilitate, where possible, re-entry into employment or placement on manpower programmes, or otherwise to provide guidance or assistance which would enhance individuals' future prospects in the labour market. If the outcome is viewed purely in terms of direct reintegration into employment, the results cannot be described as altogether impressive. Of the recommendations made in respect of long-term unemployed individuals, only about one in five related to instances where direct placement in employment was considered appropriate. Many of those who left the register did so because of the exercise of controls (e.g. failing to sign the register, non-attendance at interviews) and sizeable numbers were transferred to health-insurance lists or entered pre-retirement schemes. There was criticism that the initiative was a 'purely statistical adjustment' designed to reduce the numbers of registered long-term unemployed. It did, however, contribute to a better understanding of the problem which led to the subsequent introduction of a systematic assessment system for all unemployed entering their fourth and thirteenth weeks of unemployment (see Auer 1985).

Other countries have also utilized more intensive placement and counselling activities to help the long-term unemployed. The 'Ten Per Cent Initiative' introduced in the United Kingdom in 1981/2 is a further example, so named because it involved the allocation of 10 per cent of the resources of the Public Employment Service to helping those out of work for over a year. Evaluative work indicates that those who received additional assistance did subsequently fare better in the labour market when compared with the experience of similar unemployed persons who were accorded what one could describe as 'normal' assistance. However, while the results did signal an improvement, comparatively few of the specially targeted long-term unemployed persons (somewhat less than 20 per cent) actually regained work - rather similar to the results obtained in France in 1982/3. Even though it is too early yet to engage in evaluations, the more comprehensive Job Start programme introduced in the United Kingdom in 1986 (in which all long-term unemployed persons are offered counselling with a view to facilitating their re-entry into employment, or placement on a manpower programme) appears to have had more of an impact, at least if the position is viewed globally in terms of the decrease in long-term unemployment which has occurred in the United Kingdom over the past year or so.

Basically, however, it is very difficult to achieve the direct reintegration of the long-term unemployed into employment by means of placement and guidance activities alone. The disadvantages they suffer are formidable, both those of a personal nature such as lack of skill and demotivation, and those arising from the attitudes of employers who do not view the long-term unemployed as an attractive prospect when compared with the multitudes of other job applicants now thronging the labour market.

The foregoing comments in effect emphasize the disadvantages in attempting to utilize the second of the above-mentioned measures, employment subsidies in combating long-term unemployment. Since the selection of personnel lies ultimately with the employer, it is inevitable that the better equipped unemployed will tend to be taken on, even when employers are encouraged to do otherwise by means of special incentives. The Employment Incentive Scheme (EIS) in Ireland (which is a marginal stock subsidy-type scheme) provides a notable example of this. The number of long-term unemployed persons accommodated on this programme is minimal when viewed in terms of their share of total unemployment, even though they attract a weekly subsidy which is twice that paid in respect of other unemployed workers.

Generally speaking, firm-based employment subsidy schemes appear to have lost much of their earlier appeal, mainly because of disappointing results in the context of promoting employment growth. With few exceptions these programmes have touched only a small proportion of the unemployed (and a minimal proportion of the long-term unemployed). Furthermore, the influence of 'deadweight' and 'displacement' effects has raised serious questions as to whether the apparent employment results achieved are really meaningful in any real or net sense. One might suggest, however, that, while such programmes may not have proved effective in promoting employment creation, the potential of these schemes in relation to altering the structure of recruitment flows has not been fully exploited. There has been a lack of clarity in relation to objectives, particularly when governments have simultaneously tried to achieve the twin aims of promoting equity and employment growth. As a result, in many countries these measures have been subjected to almost continuous alteration and restructuring, frequently in an ad hoc manner without reference to an overall guiding strategy. There are indications that employment-subsidy schemes have achieved most in those countries where the equity objective was predominant - in Denmark, for example, where the Job Offer Scheme is confined solely to the long-term unemployed, and in Australia where their Job Start Scheme (not to be confused with the UK scheme of the same name) and its forerunners have always been designed as recruitment-flow subsidies targeted at disadvantaged groups (i.e. without any conditions requiring firms to record a net increase in employment).

With regard to training programmes, the reality is that

few long-term unemployed have been accommodated on what one might describe as conventional or mainstream training courses since these normally require minimum levels of competence which many long-term unemployed do not possess. Heretofore, the objectives of these training programmes have been in accordance with efficiency-related criteria as already discussed. In this context many countries have responded to the problem of the disadvantaged unemployed by introducing special reorientation programmes for groups such as ethnic minorities, unemployed persons from localities of high unemployment, young people experiencing difficulty in gaining a foothold in the labour market, women facing problems of reintegration, etc. Even though such schemes may have accommodated sizeable numbers of long-term and other unemployed persons, they are limited in many respects. They aim principally at improving motivation, life skills, and job search practices and at providing general information about the operation of the labour market. They may involve a certain amount of skill instruction, but not to the extent that the participants can subsequently exercise a particular skill or occupation. Such schemes may, however, form a bridge to more substantial skill-training programmes or to other manpower activities. Typically these programmes are of a short-term nature or, if they are longer in terms of duration, and thus more comprehensive in regard to content, they tend to cover relatively small numbers.

One might, therefore, summarize the training response to the long-unemployment problem as being not only 'limited' but also 'distinct'. The long-term unemployed and other disadvantaged groups tend to be treated separately from mainstream activities. This, of course, increases the risk of 'branding' the individuals involved which can be disadvantageous when programme participants subsequently try to re-enter the labour market. However, it can also be argued that, because of the nature of their disadvantages, the long-term unemployed really need special and distinct treatment. Thus their inclusion on conventional training programmes is frequently ruled out from the point of view of rendering effective help to the long-term unemployed themselves.

There is also, of course, the question of the disadvantageous effects that their inclusion would have on the overall effectiveness of these schemes. This is a factor

43

of some importance in the context of maintaining a national skill-training initiative in a competitive environment characterized by rapid technological change. If therefore such programmes are to be altered to accommodate greater numbers of long-term unemployed and other disadvantaged individuals, then any such restructuring will have to be gradual, retaining on the one hand the necessary elements of a strategic training plan, while on the other rendering more constructive help to the groups in question.

Finally, let us consider initiatives designed to alleviate long-term unemployment by means of direct employment creation. Initiatives of this kind are usually mounted in the form of special employment projects which involve the engagement, on a temporary basis, of previously unemployed persons on separate or clearly identified schemes designed to undertake particular tasks or to provide specific services. Such projects are usually organized as part of a comprehensive national programme which lays down guidelines within which the various schemes must operate if they are to be eligible for the funding provided. Many of the projects are sponsored by local authorities or municipalities and therefore a high proportion of the activities tend to relate to unskilled work in the construction and environmental areas. However, in recent years there has been a growing involvement in these activities of social and charitable bodies. As a result the activities covered have tended to broaden to encompass many social aspects such as youth and recreational activities, helping the elderly, mothers with young children, or even the unemployed themselves. There is usually a stipulation that the work provided through such schemes is additional in the sense that it would not otherwise be carried out via normal labour-market activities.

When it became clear from about 1982/3 onwards that long-term unemployment was likely to be an enduring and intractable problem, governments began to rely more on direct employment-creation methods in attempting to combat the problem. One can identify this period as one during which many such measures were initiated, or existing employment creation programmes greatly expanded - such as the Community Programme in the United Kingdom, the 'Plough Back' scheme in the Netherlands, the Social Employment Scheme in Ireland, the Community Employment Programme in Australia and the Troisième Circuit (Third Sector) programme in Belgium. The last-mentioned Belgian scheme is unique in that it provides employment of an indefinite duration. The public-sector element of the Danish Job Offer Scheme would also fall under this heading, even though the temporary employment involved is provided via existing public-service (local authorities) structures.

The emergence of direct employment schemes on a significant scale arose not only from a specific desire to help the long-term unemployed. It also represented an attempt to address the paradox whereby on the one hand there is manifest evidence of societal and environmental needs not being met, and on the other hand a vast reservoir of unutilized human resources in the form of massive unemployment. There are however obstacles in the way of marshalling these same resources in order to meet the unfulfilled needs referred to. In the context of the market economy this has to be done in a way that avoids adverse effects in the job market, such as displacement and the creation of unfair low-wage employment. Thus the circumstances in which direct employment-creation projects can be carried out are often limited, with the result that many of them are of marginal utility (at least when measured in conventional economic terms) and outside the realm of what one might describe as conventional work. This in turn raises questions as to whether such programmes, as currently constituted, tend to reinforce trends towards greater segmentation in the labour market. For the unskilled and disadvantaged, however, who form the great majority of the long-term unemployed, such schemes provide the only available lifeline in maintaining a tenuous connection with the labour market. Direct employment-creation measures also have the advantage that the provision of support for the long-term unemployed is guaranteed, at least up to certain limits depending on budgetary and other constraints. The other measures discussed so far can at best facilitate only modest proportions of actual jobs or substantial training for the long-term unemployed.

## CONCLUDING REMARKS

How then might one summarize the position? The first major aspect to be recognized is that, in the context of the current labour-market situation, the long-term unemployment problem will not disappear of its own accord even in the event of substantial economic and employment growth. Its resolution will, therefore, continue to require special measures. To allow the situation to drift would be tantamount to consigning large numbers of long-term unemployed to near permanent social oblivion. In this regard it must be borne in mind that those now at risk are not just hard-core unemployed whose work-seeking intentions may be suspect, or difficult-to-place older workers who have always encountered problems in regaining employment. Long-term unemployment now extends across a great many groups and reaches well down into what are termed the 'prime age' and youth categories. These remarks are not meant to imply, however, that countries have been impervious to the position. My earlier comments illustrate quite clearly that a number of countries have embarked on substantial initiatives aimed primarily at easing the plight of the long-term unemployed. If any criticisms are to be made it is to the effect that these approaches have lacked an essential comprehensiveness. Individual countries have tended to invest heavily in one or other form of measure, while the very nature of the long-term unemployment problem is such that the response to it needs to embrace a range of different remedial aspects.

The kind of approach which I am attempting to articulate is perhaps best summarized in the previously mentioned OECD report on long-term unemployment. The strategy outlined in that report involves the following elements:

> The adoption of a targeting approach concentrated heavily on the long-term unemployed, with particular provision to cater for adults and those out of work for especially long periods.

> National placement services have a vital role to play in identifying and reaching the long-term unemployed, in striving to achieve their direct reintegration into employment where this is possible, and in guiding them to the most appropriate forms of manpower support. In this context a further matter of prime importance is the 'preventive' aspect, i.e. targeting assistance on those unemployed who are more likely to lapse into long-term unemployment.

> Encouraging, where appropriate, the private sector to reabsorb the long-term unemployed through the use of

enterprise-based recruitment (as distinct from marginal-stock) subsidies. Some countries, may, however, wish to try to achieve this objective by somewhat different means, e.g. by channelling support payments directly to the long-term unemployed.

The training of the long-term unemployed and other disadvantaged groups constitutes one of the most serious challenges facing training authorities. While it is recognized that a balance has to be struck between the provision of strategic skill training and programmes involving greater remedial content aimed at disadvantaged groups, the long-term unemployment problem is now so deep rooted and is of such a scale that serious consideration will have to be given to achieving a gradual reorientation of all training programmes so that they provide greater assistance to the disadvantaged unemployed.

Until such time as a more broadly-based range of measures has been developed, it is only through direct employment-creation schemes that many of the more disadvantaged long-term unemployed can be helped. Efforts should, therefore, be made to enhance the forms of work on offer so that the participants can subsequently compete more effectively in the labour market or be able to take advantage of the opportunities arising from other manpower schemes.

There are a number of comments which are relevant in relation to the foregoing suggestions. In the first place, the fact of financial constraints has to be recognized. It is not envisaged that each element of the proposed strategy can be pursued at will. Within the context of overall budgetary limits the emphasis given to each element would depend on how governments perceive the nature of long-term unemployment in their respective countries. However, irrespective of degree, each element has a role to play in the context of a coherent approach which takes account of current labour-market conditions. The emphasis on recruitment subsidies, for example, is aimed at catering for the better-equipped long-term unemployed, thus allowing schemes such as employment-creation projects more scope in assisting the disadvantaged. In the context of achieving greater coherence and co-ordination the report also expresses the view that countries which have one all-embracing manpower agency are in a more advantageous position.

It should also be noted that the targeting provisions mentioned (i.e. in favour of the adult long-term unemployed) would require fundamental changes in the provisions governing the European Social Fund, in particular the current stipulation which requires that 75 per cent of the Fund's advances must be confined to youth-oriented schemes. It must be remembered that in Northern European countries at any rate, over three-quarters of the long-term unemployed are aged 25 years or over.

The nature of the above-mentioned recommendations was significantly influenced by the problems presented by the prior existence of a large bloc of long-term unemployed persons. What, however, are the longer-term prospects? Are many countries, for example, faced with a period when the continuing pace of technological change and labour-market restructuring has the potential to, as it were, create a 'permanent' long-term unemployment problem - unless mechanisms are evolved which re-equip displaced workers and enable them to be reabsorbed into employment before demotivation and deterioration in human capital sets in? It was issues such as these which prompted the suggestion that in the longer term a reorientation of training programmes is needed which would give much greater priority to assisting groups such as the long-term unemployed. This, however, raises wider issues. It is possible that such provision could be made available only in the context of a comprehensive manpower-support system which would embrace both financial compensation for unemployed persons and an integrated set of aids including training, retraining, further education, and, where necessary, the provision of employment. Such a system, while it might provide the ultimate deterrent to long-term unemployment, would, of course, have considerable cost implications and could be difficult to sustain, particularly in periods of sluggish economic growth.

Some would consider that such an approach would impose an undue degree of further regulation on the labour market, when the general tendency now is to loosen up the system and render it more flexible. However, the notion of a more flexible labour market is not necessarily incompatible with a comprehensive manpower-support system. More flexible labour-market arrangements could convey considerable benefits in the form of higher levels of employment, even if at the expense of a more volatile employment scene with greater flows into and out of jobs. This to my mind renders it all the more necessary to have a manpower-support system which would mitigate and offset many of the resultant ill effects by aiding the reintegration of displaced workers before demotivation and deterioration in human capital occur.

## REFERENCES

1    Much of the content of this chapter has been drawn from an OECD Report with which the author was associated: OCED, Directorate for Manpower, Social Affairs and Education (1988) Measures to Assist the Long-Term Unemployed - a Review of Recent Experiences in OECD Countries, Paris: OECD.

2    The total 1985 Community unemployment level, when measured in Labour Force Survey terms, for the twelve countries in question was 14.8 million. However, in respect of some 1 million of these either no information on duration of job search was available, or it was not of particular relevance, as in the case of persons on lay-off, etc.

Auer, P. (1984) Reintegration of the Long-Term Unemployed: An Overview of Public Programmes in Eight Community Countries, Report prepared for the EEC Commission, Berlin: International Institute of Management.

EUROSTAT (various issues) Community Labour Force Survey, Luxemburg.

OECD (1964) Recommendations of the OECD Council on Manpower Policy as a Means for the Promotion of Economic Growth, Paris: OECD.

OECD (1976) Recommendations of the OECD Council on a General Employment and Manpower Policy, Paris: OECD.

OECD (1986) Labour Market Flexibility - the Current Debate: A Technical Report, Paris: OECD.

OECD (1988) Measures to Assist the Long-Term Unemployed - A Review of Recent Experiences in OECD Countries, Paris: OECD.

OECD (various issues) Employment Outlook, Paris: OECD.

Sinfield, A. (1968) The Long-Term Unemployed, Paris:
    OECD.

**Chapter Three**

## THE EUROPEAN COMMISSION AND LONG-TERM UNEMPLOYMENT

Derrick Johnstone

Long-term unemployment policy is a tale of good intentions - albeit relatively unfulfilled - on the part of the European Commission. This chapter traces the development of long-term unemployment as a policy issue in the European Community to the point, at the beginning of 1988, of great speculation about the future of the European Social Fund (ESF), the main source of European funding for initiatives to counter long-term unemployment. It concentrates on two Commission policy documents - the Action to Combat Long-Term Unemployment (1984b) and the Memorandum (1987d) on the same subject. The concluding section sets the future of financial support through the ESF in the context of the passing of the Single European Act and the consequent reform of the 'structural funds' (the European Regional Development Fund and the European Agricultural Guidance and Guarantee Fund - Guidance Section, as well as the ESF). An underlying theme is the extent to which the European Commission can and does influence the employment policies of member states.

## THE 1984 COMMUNICATION

The formal interest of the European Commission in long-term unemployment can be traced back to a meeting of the Joint Council of Finance and Employment Ministers on 16 November 1982. The Joint Council asked the European Commission to undertake a study of long-term unemployment and make proposals for remedial action. In September 1984, as a consequence of this initiative, the

51

Commission published the Communication to the Council and the Standing Employment Committee entitled Action to Combat Long-Term Unemployment (1984b). This document provided a thorough review of long-term unemployment: of the scale and nature of the problem, of the characteristics of those individuals affected, and of the associated social and economic costs. The language of the Communication conveyed an indignation about the waste of human resources that large numbers of long-term unemployed constitute. It was notably critical of actions by member states in pointing out that:

> Overall, there does not appear to have been much systematic thinking about the scope and scale of the measures, about their coherence with respect to overall economic and social objectives, or about longer-term strategies. (p.8)

The review of national measures highlighted particular weaknesses, such as the extent to which the style and content of education and training programmes were ill matched to the needs of the unemployed, and also a lack of effort to encourage unemployed people to take advantage of these educational and training opportunities. The Commission recognized that 'almost invariably' successful programmes for the long-term unemployed had been based firmly in local communities, with close working relationships between clients and the wide range of bodies that can assist: public authorities, employers, and trade unions, the churches and community groups, educational and training institutions and so on. Repeatedly over the past few years, documents of the European Commission have stressed the need for locally based partnerships and strategies and have criticized - implicitly or explicitly - member states for a lack of local flexibility in many of their employment and business-development programmes.

As far as policy initiatives by the European Community itself were concerned, the Communication drew attention to commitments expressed in other policy documents: in the fields of vocational training (CEC 1982a), youth employment (CEC 1983a) and local employment initiatives (CEC 1983b). In vocational training, priority was to be given to young people and the long-term unemployed, especially those who lack the basic knowledge and skills usually required to take part in training programmes (Council Resolution, 11.7.83 OJ

No C193, 20.7.83). Local employment initiatives, as defined by the European Community, encompass a range of non-conventional forms of enterprise: co-operatives, community businesses, small businesses set up by redundant steel workers, and so on. Support for local employment initiatives has been expected to help unemployed people, in particular those lacking capital and business experience but willing to make an effort to create work for themselves (Johnstone 1986). The Commission had also initiated a debate on social-security issues relating to long-term unemployment (CEC 1982b) and had given priority to the needs of such unemployed people within the framework of the Second Poverty Programme (CEC 1984a) (see the chapter by John Benington).

By far the most important means of European Community financial support has been through the European Social Fund. Support, for example, for local employment initiatives under the European Regional Development Fund has been marginal. The significance of long-term unemployment was recognized formally in the ESF guidelines introduced for 1984-6, in stipulating that assistance for recruitment aids and 'public-utility' job creation was to be limited to job seekers under 25 and the long-term unemployed. While such assistance has been restricted to priority areas, scope for actions to help the long-term unemployed was opened up in the guidelines, initially with no regional limitation, and covering 'operations forming part of local initiatives aimed at creating additional jobs or the socio-occupational integration of categories of person disadvantaged in relation to employment'. The Innovatory Projects section of the ESF also extended possibilities; proposals were expected to introduce new forms of training, in content and method or organization, and lead to the development of policy and practice.

There has also been a trend within European Community policy-making towards a greater concentration of assistance and the linkage of structural funds. Long-term unemployment has been cited as one of the criteria to be taken into account in determining the location of such concerted measures. However, weaknesses in statistics on long-term unemployment have made it difficult to apply this criterion in practice. The problem here is that long-term unemployment has been defined in different ways in different countries: there are variations in the period considered 'long-term', in the eligibility of those out of work

**Table 3.1** Recommendations contained in Action to Combat Long-Term Unemployment

---

### Action by member states

1 Improve information on long-term unemployment (including the characteristics and circumstances of those affected).
2 Provide counselling and training for workers about to be made redundant.
3 Ensure that employment services are structured, organized, and staffed to provide more personal contact to unemployed people.
4 Collaborate with other bodies in providing temporary work initiatives (which should involve an element of education/training).
5 Fund local centres for unemployed people.
6 Encourage greater awareness of the problem of long-term unemployment and possible solutions.

### Action by employers

1 Provide appropriate support for workers facing redundancy.
2 Develop youth-training and employment policies.
3 Take up inducements to recruit long-term unemployed people.
4 Review practice of setting restrictive age limits on recruitment to certain jobs.
5 Sponsor local economic initiatives.

### Action by trade unions

1 Facilitate the continued participation in trade-union activities of out-of-work members.
2 Negotiate more flexible terms of recruitment which could encourage employers to offer more vacancies to the long-term unemployed.
3 Support and encourage the creation of unemployed workers' centres.

---

Source: CEM (84) 484 final

for unemployment benefit, in employment measures which hide the true scale of unemployment (when taken to mean the number of people without a job who would take one if a suitable job were available). Although some standardization is possible, available statistics on long-term unemployment at a country and regional level cannot be compared across the European Community.

It is worth restating a further part of the Communication's conclusion, for it remains pertinent today:

> there is reason to fear that the long-term unemployed may be among the last to benefit from any upturn in employment since employers will tend to draw on those more recently unemployed when they begin to hire again. All efforts should clearly be made to ensure that this sombre forecast does not become reality, and policy actions to combat long-term unemployment need to be both strengthened and better targeted if some measure of success is to be achieved over the medium-term.

The European Commission went on to call for such strengthening by actions of national, regional and local governments, and by employers and trade unions (the 'social partners'), to promote economic development, to anticipate and respond better to changes in the economy and to ensure the best use of European Community support. A long list of practical recommendations was put forward for action by member states, by employers and trade unions and by the European Community (see table 3.1).

The European Commission itself proposed taking steps to achieve a better understanding of the nature and extent of long-term unemployment, to support its recommendations for action by member states and social partners, and to identify and disseminate successful measures and practices throughout the Community. A means of achieving the latter objective was MISEP, the Mutual Information System on Employment Policies, set up in 1984 by the European Commission and concentrating on national policy measures.

The European Commission had also called for a 'broad policy reappraisal' to be undertaken at Community level 'in order to develop a more coherent medium-term approach', highlighting not just new measures and initiatives to promote job creation but also the need for an integrated and coherent system of income support, new policies on retirement, further help for disadvantaged young people, more flexible patterns of working, and training and education more appropriate to the needs of unemployed people. The idea of a social guarantee was floated, similar to that proposed for young people by the European Commission, whereby unemployed people would be offered placement in a training or employment scheme when they had been out of work for twelve months. This 'broad policy reappraisal' has not taken place, although elements of it have been subsumed within other policy deliberations. When the attitudes of member states to the proposal of a social guarantee were sounded out, the response was cool. They were concerned that such a guarantee would not improve access to jobs and would draw funds away from training for young people and for technology; moreover, in most countries access to training is regarded as an individual right and responsibility (Mawson 1986).

Generally speaking, the Communication was not very well received by the member states. It was politically premature, raising long-term unemployment as an issue at a stage before it emerged as a priority subject on national policy agendas. Its stance and language were not in keeping with the philosophy of right-of-centre governments in the European Community. But it can be argued that in setting out the nature of the problem, and in synthesizing and illustrating policy measures in different countries, the Communication helped EC countries to learn from one another, a process that has been further developed through MISEP.

## ACTION SINCE THE 1984 COMMUNICATION

Action by the European Community since the 1984 Communication can be divided into two categories: Community funding and research and information exchange.

### Community funding

Long-term unemployment has continued to feature within the European Social Fund guidelines. In the guidelines for 1988-90 support was available for projects providing vocational training for the long-term unemployed, 'geared to their needs and including motivation and guidance and offering substantial prospects of employment', but only in the priority areas (Guideline 4.5). On the one hand, support for 'recruitment or setting up of the long-term unemployed to additional jobs of indeterminate duration or placement in additional jobs of at least six months which fulfil a public need' is available in the UK only in Northern Ireland. On the other, support continues to be available for innovatory projects and for 'local initiatives'. The latter are available in priority regions only for the over-25s, with no restriction for such projects employing young people.

Twenty-five per cent of expenditure under the European Social Fund was devoted to long-term unemployment in 1986; two-thirds went to young people under 25, one-third to over-25s. It was expected that this figure would rise to 40 per cent in 1987. About half of this expenditure has been in the form of recruitment subsidies. Innovatory projects aimed at long-term unemployed people accounted for 1.5 per cent of ESF expenditure (CEC 1987d). At the same time it is important to note that the number of long-term unemployed adults as a proportion of all people assisted under the European Social Fund is much lower in the UK than the Community average (3.2 per cent in the UK compared to an average of 10.5 per cent in 1987) or particular countries such as France (16 per cent) and Spain (19 per cent). This situation reflects the high proportion of the ESF going to the UK and the bias within this figure towards support for young people. Also, the UK government has not promoted recruitment subsidies for unemployed adults and has lagged behind France in the development of monitoring and counselling services for long-term unemployed people.

A note of criticism is warranted. Much of this ESF expenditure has financed national employment programmes and has been treated by member states as part of their 'rebate' for their contribution to the European Community budget. It has not led in such cases to additional provision for long-term unemployed people, and no recognition has been made publicly of the source of funding. ESF support for innovatory projects can also be criticized, even when the projects are sponsored by agencies other than national governments. As mentioned above, such projects are expected to introduce new forms of training, in content, method or organization, and to lead to the development of policy and practice. However, mechanisms have never been put in place for a systematic appraisal of such pilot projects and for dissemination of their results and lessons.

## Research and information exchange

The European Commission has sponsored a variety of research and information activities relating to long-term unemployment. These activities have included:

> continued funding of MISEP, which has become established as a comprehensive database on national employment measures, producing country reports, regular bulletins etc.; (1)
> research into activities for the unemployed through programmes managed by the European Foundation for Living and Working Conditions on the role of vocational training in assisting the long-term unemployed (by Quaternaire Education, Paris, and the Economic and Social Research Institute, Dublin - for details see Mawson 1986), and into national employment measures introduced since the 1984 Resolution (by the Institute of Manpower Studies);
> publications, such as 'Long-Term Unemployment: The Way Out?', researched and published by ELISE (the European Information Network on Local Employment Initiatives); (2)
> international seminars and conferences, including 'The re-entry of long-term unemployed people into the labour market', organized by Canal-Emploi, Liège, in September 1985 (for details see Morgan 1986) and 'Unemployment: information and intervention', organized by Strathclyde Regional Council in October 1987;
> associated research programmes such as the Action-Research Programme on Local Labour Market Development (see the chapter by Haris Martinos), relevant in the context of preventative measures to counter long-term unemployment.

MISEP has been particularly important not only in promoting information exchange between member states but also in enabling the European Commission to monitor member state policies and their broader context.

## ACTION BY MEMBER STATES

At the time of the 1984 Communication, there were relatively few initiatives of member states aimed at tackling long-term unemployment. Political priority was accorded to measures to counter youth unemployment. However, since then, there has been a growing recognition that the problem of long-term unemployment is not going to disappear of its own accord, and a realization that such unemployment is now the average condition, not the exception. In 1985 over 50 per cent of unemployed people in the European Community had been without work for over a year (a figure ranging from 32 per cent in Denmark to 68 per cent in Belgium). Moreover, this proportion had grown markedly in the previous four years, especially in West Germany and the Netherlands.

The 'problem' of long-term unemployment has been associated with national variations in responses. Even so, all member states now operate at least one measure aimed at long-term unemployed people. These policy measures can be categorized broadly into three groups: direct public job-creation programmes (generally providing work of benefit to local communities); training and retraining courses; and recruitment subsidies. In addition, long-term unemployed people have access to general employment and training programmes, although they may not have relevant experience, qualifications, or motivation to take advantage of such opportunities.

There was a shift of emphasis after 1984 in member states away from efforts to stimulate demand for labour through job-creation measures to efforts to improve the quality of labour supply through vocational guidance and training. The extent to which this shift has occurred, specifically targeted at the long-term unemployed, was limited by 1987, and small in comparison to the scale of the numbers of people involved. There are signs of a trend towards more comprehensive approaches, with an emphasis on monitoring, guidance, and various options such as a place on a training course if no suitable job exists.

## THE 1987 MEMORANDUM

At the meeting of the European Council in June 1986 the heads of government expressed their concern over the growth of long-term unemployment. The Presidency - provided by the Netherlands - indicated in its conclusion that:

> With a view to supporting a convergent European policy aiming at the return of the long-term unemployed to the labour market, the European Council welcomed the Commission's proposals on exchanging information on successful national experiences, on conducting pilot actions under the European Social Fund, and on co-ordinating actions under the Community structural instruments in regions in need of restructuring. (quoted in CEC 1987b)

Following this declaration, in December 1986 the European Council identified long-term unemployed people as requiring particular help, as part of its resolution on an Action Programme for Employment Growth.

This Action Programme was the outcome of an initiative promoted by the British, Irish, and Italian governments during the British Presidency, and was entitled 'Strategy for the labour market - employment growth in the 1990s' (Employment Gazette 1986). In addition to support for the long-term unemployed, the Action Programme for Employment Growth covered measures to promote enterprise and employment (through deregulation, training, information, and advice for small businesses, etc.), to encourage more flexible employment patterns and conditions of work, and to develop vocational training to help meet the changing requirements of employers and rapid technological advances (CEC 1987f).

In May 1987, these resolutions were followed by the Memorandum from the Commission on Action to Combat Long-Term Unemployment (CEC 1987d). It was intended to stimulate a further 'preliminary' discussion, in particular on measures proposed by the European Commission (described below), and served, in part, as an interim report on the implementation of the 1984 resolution. The conclusions and recommendations of the Memorandum were based on analysis of a questionnaire sent to member state governments, on information derived from MISEP and on research carried out both for the European Commission and for other bodies.

The Memorandum put forward a series of propositions to reinforce measures to tackle long-term unemployment:

## Target reduction in long-term unemployment

The European Commission advocated the objective of reducing the proportion of long-term unemployed in the total, specifically from 50 per cent to 30 per cent by 1990. In addition, it proposed that those out of work for twelve months or more should be guaranteed at least a 'certain level' of counselling and assessment and, ideally, assistance in seeking available jobs or a place on an employment or training programme.

## Reintegration of the long-term unemployed

The Memorandum stressed that measures to assist long-term unemployed people must focus on making the individual as attractive to employers as newly unemployed workers or new entrants to the labour market. This reintegration should be achieved through:

skills training and retraining courses (including basic work skills, literacy/numeracy and job-specific skills);

recruitment incentives to employers;

integrated schemes offering a package of counselling, basic skills training, and recruitment support.

The European Commission recognized that approaches need to take account of the differing circumstances of local labour markets and of the needs and capabilities of the individual. It noted that those best qualified and least in need of assistance among the long-term unemployed will tend to be those who benefit from recruitment subsidies, and that direct job-creation programmes are likely to be needed to fill the gap for those people most at a disadvantage in competing for jobs. The European Commission also called on member states to ensure that business advice and guidance were available to unemployed people who wished to create jobs for themselves.

61

**Preventative measures**

Advice from the European Commission about preventative action fell into two categories.

A monitoring system which 'triggers off' support for individuals at risk of prolonged unemployment, both those newly unemployed (especially those displaced by industrial restructuring) and those out of work for long periods and particularly at risk of losing all contact with the labour market. This system may take the form of further help in job search or a training opportunity.

'Strategies to promote local labour-market development', namely local strategies which combine business development, training, and employment measures. The European Commission argued that:

Member states should help to develop support for structures and training for new entrepreneurs in local communities and encourage local-level initiatives to develop new employment opportunities.

The European Commission also argued that any longer-term strategy must include action to improve levels of educational attainment, basic training, and access to continuing training throughout a working life.

To underpin these preventative measures, the European Commission saw as a prerequisite improved monitoring of changes in the labour market, at both local and national levels.

**Reorganization of national employment services**

There was an implicit concern of the European Commission that many national employment services had lacked the political direction to focus on long-term unemployed people as a priority. Rather their focus had been on helping employers fill job vacancies with the best applicants. The European Commission called for a political lead aimed at ensuring that all national employment services were structured and equipped to identify and follow up all people out of work for twelve months or more. Publicity campaigns were suggested to draw to the attention of both jobseekers and employers the range of employment and training programmes available.

## PROPOSALS OF THE EUROPEAN COMMISSION

The European Commission proposed to implement a four-part programme of action consisting of promotional activities to show how long-term unemployment can be tackled, exchange of national experiences and evaluation, improved statistical information, and Community financial support.

### Promotional activities

Promotional material was needed, designed to increase awareness and understanding of the problem of long-term unemployment and to show what could be done to tackle the problem. Such material should be aimed at the social partners, especially employers. It would draw on existing sources of information such as ELISE, MISEP, and CEDEFOP (the European Centre for the Development of Vocational Training in Berlin) and on additional information provided through research and case-study material.

### Exchanges

The European Commission would intensify and extend its current activities to promote the exchange of information between member states, particularly on implementation and evaluation. This exchange would be achieved through MISEP, the existing system of exchange of national officials and meetings of top civil servants responsible for employment policy and services. (3)

### Improvement of statistical information

Further improvements were needed in statistics on long-term unemployment, both in developing comparable bases and in making full use of Community Labour Force Surveys (carried out annually to provide comparable labour-market data throughout the European Community). (4)

## Community financial support

Financial support for training and recruitment would continue to be provided through the European Social Fund. (The possibility has been floated that there could be a programme of demonstration projects on long-term unemployment within the new framework of the Social Fund.)

## THE EUROPEAN COMMISSION'S SEARCH FOR CONSENSUS

A series of points for discussion was put forward to help establish consensus about the nature of the problem of long-term unemployment and about appropriate policies and measures. It was believed that, once such consensus was established, efforts to tackle the problem would become more coherent and effective.

The European Commission sought positive responses from the Standing Committee on Employment and from the European Council. It looked for agreement on its proposals for the target reduction in long-term unemployment, for a programme of demonstration projects under the European Social Fund, for enhancements of the MISEP system, and for reforms where necessary in national employment services. It also sought views on whether the structural funds should do more to encourage experimentation and innovative projects or to support existing measures adopted by member states. Setting aside a higher proportion of funds for innovatory or demonstration projects was one way in which the European Community could hope to exert greater influence on the development of policy and practice in member states. Otherwise, without this carrot, the European Community must rely largely on exhortation, lacking the stick with which to beat the member states.

The outcome of the meetings of the Standing Committee (5 November 1987) and of the European Council (1 December 1987) did not go as far as the European Commission had hoped. The European Council recognized that, while economic growth was a necessary condition for any reduction in unemployment, specific measures were necessary to tackle long-term unemployment (Council Resolution, 1.12.87, OJ No C335/1). It said that such measures should include: counselling and regular follow-up of unemployed people; assistance with job search; training and work experience; support for business start-ups; and recruitment aids.

Implicitly these measures were seen to be the responsibility of individual member states. No reference was made in the European Council's conclusions to the proposal for the programme of demonstration projects, nor to the role of the structural funds. Nor were there any references to the links between macro-economic and labour-market policies, the role of education and training, and the social consequences of unemployment, though these points had been made in the Memorandum and supported by the Standing Committee. Moreover, no commitment was given to the proposed target reduction in long-term unemployment.

What the European Council did agree was to invite the European Commission to prepare a 'programme of action' comprising:

identification and dissemination at Community level of details of successful national programmes;

enhancement of existing systems of information exchange (MISEP, ELISE), in particular asking MISEP to assemble more information on the design, effectiveness, and costs of national measures, with details also requested about the experiences of other OECD countries;

improvement of statistics on long-term unemployment.

Whilst the European Council did not go as far in its conclusions as was desired, the significant outcome was that tackling long-term unemployment was reaffirmed as a priority of the European Community.

## Information-exchange proposals

After the meeting of the European Council in December 1987 a proposal emerged to act on the recommendation about information exchange. The idea was to launch the pilot stage of a project modelled to some extent on EuroTechneT, an existing network of demonstration projects located throughout the European Community. This latter network is funded under the European Commission's Work Programme (1986-8) on Vocational Training and New Information Technologies (CEC 1985). EuroTechneT was intended:

> to use the developing 'good practice' within the network's projects to influence the quality and effectiveness of as many relevant training schemes and projects as possible;

> to promote and contribute to the supply of 'tested-and-proven' reference material about training design and delivery;

> to ensure the effective evaluation of projects in the network;

> to promote exchange of experience between project managers in the various member states.

It operates through regular meetings of network projects in each country, through the publication of a directory of projects, and by financing visits of project managers to other projects. EuroTechneT does not otherwise provide funds directly to participating projects, although a number are funded by the European Social Fund. The European Commission is using the network to provide information and ideas to contribute on research and development and training policies (Harrison 1986). Similarly the European Commission would be looking to this network of long-term unemployment projects to help to identify successful, proven approaches which could be replicated elsewhere and financed under the new guidelines for the reformed European Social Fund (see Chapter 4). This development would help to overcome one of the current criticisms of the administration of the European Social Fund: that no effort is being made to identify and disseminate the lessons of innovatory projects to a wide audience. The European Commission intends also that the network projects will be well publicized to demonstrate the potential of new measures and initiatives.

## THE SINGLE EUROPEAN ACT AND REFORM OF THE STRUCTURAL FUNDS

Of great significance for West European integration is the passing of the Single European Act on 1 July 1987 and the consequent reforms of the structural funds. The Single European Act was intended to bring about institutional changes and set new objectives for the European Community, notably the completion of the 'internal market' by 1992 and the 'strengthening of economic and social cohesion'. This latter objective involved, in essence, efforts to reduce inequalities and imbalances between regions and between people. The new Act prescribed that changes should be made to the structure and operational rules of the European Social Fund, the European Regional Development Fund, and the European Agricultural Guidance and Guarantee Fund (Guidance Section), in order to 'clarify and rationalize' their tasks, increase their efficiency, and co-ordinate their activities not only between the Funds themselves but also with the other financial instruments of the European Community, such as the European Investment Bank.

This reform focused on five objectives, one of which was combating long-term unemployment. The others were promoting backward regions, helping declining industrial regions to adapt, integrating young people into the labour market, and encouraging rural development and agricultural reform. Of these, the principal objective was that of enabling less favoured regions 'to catch up'. The Communication, Making a Success of the Single Act (CEC 1987b), stated that support for actions against long-term unemployment should be available irrespective of geographical criteria.

These reforms were to be implemented through the Framework Regulation (CEC 1987e) to be followed by regulations for the individual Funds. The Framework Regulation proposal restated the five priority objectives set out in the Single European Act and identified the European Social Fund as the vehicle for addressing the objective of combating long-term unemployment (referred to as 'Objective No 3'). It assigned to the Social Fund the task of providing 'support for measures, notably in the field of vocational training, aimed at: (a) securing better use of, and adapting human resources; and (b) expanding employment opportunities' (Article 3.2).

The Regulation Proposal stressed that 'Community operations shall be such as to complement corresponding national operations' (Article 4.2) and that they shall be established through 'partnerships' between the European Commission, the member state, and 'where appropriate', regional, local, or other authorities 'designated by that member state'. Such partnerships were to cover the preparation, financing, monitoring, and assessment of operations. Preference was to be given to multi-annual programmes. This change would be particularly significant for the European Social Fund, which has been administered on the basis of annual submissions, project by project.

The European Commission's philosophy was set out in 'The Commission's programme for 1987' (1987c) as follows:

Community intervention must become more and more integrated into programmes negotiated with national, regional and local government. These programmes should embrace not only aid for investment, further vocational training and other measures to encourage geographical mobility, but also aid for the development of an environment that will foster the continuation of existing activities and the introduction of new ones. The programme approach implies co-ordination between the three separate Funds, thereby allowing sufficiently specialized know-how to be developed. It should also open the door to more decentralization of the execution of operations and the diversion into technical assistance and monitoring capacity of resources at present devoted to administrative tasks. (p. 38)

It was recognized that the financial resources currently allocated to the structural funds were too modest to secure any marked reduction in regional imbalances; hence the European Commission and European Parliament pressed for a doubling of the funds available to the ESF and ERDF. At the Brussels summit of February 1988 agreement was reached that expenditure on the structural funds should increase from ECU (European Currency Unit) 7.8 billion in 1988 to ECU 13 billion by 1992. At the same time, the European Commission argued that interventions by the structural funds would have greater impact if they were more selective and concentrated. At the same summit it was agreed that the structural funds were to be more focused than previously on economically backward areas, which were to get a doubling of aid by 1992.

The European Commission was to establish general guidelines 'that set out and clarify the Community choices and criteria concerning action to combat long-term unemployment' (Article 10.1 of the Single European Act). There is to be a 'support framework' which will indicate the Community objectives; the particular objectives adopted for Community assistance; the forms of assistance; the financial plan; and the duration of assistance (Article 10.3). Member states are required to submit plans for combating long-term unemployment, to include: information on national employment policy; an indication of priority operations already under way or to be carried out; and an indication of the use that national authorities intend to make of the European Social Fund (Article 10.2). Similar requirements are stated in relation to Objective No 4, facilitating the occupational integration of young people.

## Current situation

The situation with respect to the Framework Regulation was that the future funding of the Community had to be resolved first. The key date was the European Summit Meeting held on 11 February 1988. The Framework Regulation had then to be adopted (and within one year of the passing of the Single Act, i.e. by 30 June 1988), to be followed by implementation regulations for each of the structural funds. It was hoped that the legislative reforms would be completed by 1 January 1989 with the operational changes made by the start of 1990, but the budgetary delays brought these targets into question.

What will these changes mean for local authorities wanting to develop initiatives to assist the long-term unemployed and looking to the European Community for funding? Although an answer to this question involves speculation, some pointers can be given.

Fewer local authorities are likely to have priority status under the ESF, but those that continue to do so may have access to more funds, and without threat of the 'linear reductions' applied in the past to share out funds between excessively large numbers of eligible projects.

The European Commission will be looking to support projects which involve the ESF <u>and</u> the ERDF, and, indeed, other EC measures (such as in the expanding environment field, newly given legal status under the Single European Act).

More importantly, the European Commission will be looking to fund programmes which combine the structural funds, are multi-annual in character, and are submitted through the national government. What is unclear at present is the role that local authorities - and for that matter, voluntary organizations - will play in devising and implementing these programmes. (Under the Framework Regulation Proposal the national government may 'designate' that local authorities can participate in the new 'partnership' arrangements with the Commission.)

In the meantime, it can be argued that local authorities should not hold back from taking steps to develop a strategy against long-term unemployment, as they can be secure in the knowledge that this subject will be one of the key objectives of the reformed Social Fund. However, uncertainty remains about whether or not funds will be available on any significant scale to long-term unemployment projects located outside priority areas. This latter issue is one of many being debated within the European Commission in early 1988. There is a fundamental lack of clarity in the hierarchy of the five goals embodied in the Single European Act because of the mixture of 'area-based' and 'people-based' targets. Having a strategy and thus a marshalled case will be a considerable help when opportunities for ESF funding arise and, in any case, will be relevant to obtaining funds from UK sources such as the Training Commission.

## REFERENCES

1    MISEP - Mutual Information System on Employment Policies. Its function is to provide information on employment measures operating in member states, in order to assist in the formulation of policy. Regular output includes: basic information reports (BIRs) for each member state, containing descriptions of

individual employment measures; InforMISEP newsletter, containing articles describing recent
developments in member states on employment
measures; and synoptic tables summarizing measures in
each country. The European Commission's quarterly
journal Social Europe regularly features information
generated through MISEP. Contact: European Centre
for Work and Society, Postbox 3073, NL-6202
Maastricht, The Netherlands - tel: 010 31 43 216 724.

2 ELISE (European Information Network on Local
Employment Initiatives). It was set up in 1984 with
European Commission funding to provide an information
service on local employment initiatives throughout
Europe. Working with a network of correspondents in all
member states, ELISE has built up a specialized
databank/documentation centre and provides an enquiry
service. ELISE produces a wide range of publications,
including: ELISE Feedback (quarterly bulletin prepared
in conjunction with the OECD); European Press Review
(monthly); ELISE Info-Research (monthly newsletter);
CEE News (fortnightly briefing on EC policy);
information sheets on particular topics (e.g. on projects
funded by the ESF); topical reviews (e.g. of youth
enterprise, long-term unemployment, and the role of
trade unions); yearbooks (e.g. on sources of information
on local economic development); and summaries of EC
research reports. Contact: ELISE, 38 rue Vilain XIIII,
1050 Brussels - tel: 010 322 647 2400.

3 Exchanges of employment officials. The European
Commission has a long-established programme for
exchanges between officials of the employment
services of member states, with its origins in planning
for the integration of the Community labour market.
Its emphasis has been on developing co-operation on
practical issues such as staff training, the introduction
of data processing techniques, and exploring the subject
of vocational guidance. From 1987 onwards, the study
of subjects with implications for new national or
Community employment measures has been possible.
Each year two topics are chosen for such study. In 1987
these were recent experiments intended to improve the
workings of the labour market; and transition of young
people from training to employment. Exchanges may
take the form of brief meetings or visits or six-week
projects (Silletti, 1987).

4    Long-term unemployment statistics. Comparable statistics of long-term unemployment are essential if this indicator is to be used in the definition of priority areas for Community assistance. However, available statistics are not on a standardized basis amongst member states, particularly at Level III of statistical disaggregation (regions in Scotland, counties in England). 'Long-term' unemployment is defined differently in different countries; there are variations in entitlement to unemployment benefit, and so on. The picture is blurred also by the existence of various national employment measures and, particularly in some regions, the existence of underemployment. Options to tackle the problem include an extension of the Community Labour Force Survey (costly and likely to be resisted by member states) or requesting the employment services of member states to collect the necessary standardized data (again likely to be resisted by national governments).

Aitken, P. (ed.) (1987a) A Guide to EC Organization, Operations and Funds, Glasgow: Planning Exchange Briefing Note 13.
Aitken, P. (1987b) The European Social Fund 1988-90: A Background Guide and Notes for Applicants, Glasgow: Planning Exchange Briefing Note 14.
CEC (Commission of the European Communities) (1982a) Draft Resolution of the Council Concerning Vocational Training Policies in the European Communities in the 1980s, COM (82) 637 final, Brussels: CEC.
CEC (Commission of the European Communities) (1982b) Social Security Problems - Points for Consideration, COM (82) 716 final, Brussels: CEC.
CEC (Commission of the European Communities) (1983a) The Promotion of Employment for Young People, COM (83) 211 final, Brussels: CEC.
CEC (Commission of the European Communities) (1983b) Community Action to Combat Unemployment: The Contribution of Local Employment Initiatives, COM (83) 662 final, Brussels: CEC.
CEC (Commission of the European Communities) (1984a) Proposals for a Council Decision on Specific Action to Combat Poverty, COM (84) 379 final/2, Brussels: CEC.
CEC (Commission of the European Communities) (1984b) Action to Combat Long-Term Unemployment. COM (84)

484 final, Brussels: CEC.

CEC (Commission of the European Communities) (1985) Training and the New Information Technologies: Work Programme 1985-1988, COM (85) 167, Brussels: CEC.

CEC (Commission of the European Communities) (1986a) Mutual Information System on Employment Policies: Basic Information Reports, Maastricht: European Centre for Work and Society.

CEC (Commission of the European Communities) (1986b) Fourteenth Report on the Activities of the European Social Fund, COM (86) 583 final, Brussels: CEC.

CEC (Commission of the European Communities) (1987a) 'Long-term unemployment: recent trends and developments', Social Europe, 1/87.

CEC (Commission of the European Communities) (1987b) Making a Success of the Single Act, COM (87) 100 final, Brussels: CEC.

CEC (Commission of the European Communities) (1987c) 'The Commission's programme for 1987', Bulletin of the European Communities, Supplement 1/87.

CEC (Commission of the European Communities) (1987d) Memorandum from the Commission on Action to Combat Long-Term Unemployment, COM (87) 231 final, Brussels: CEC (published also in Social Europe 3/87).

CEC (Commission of the European Communities) (1987e) Proposal for a Council Regulation on the Tasks of the Structural Funds and Their Effectiveness and on Co-ordination of Their Activities Between Themselves and with the Operation of the European Investment Bank and the Other Financial Instruments, COM (87) 376 final, Brussels: CEC.

CEC (Commission of the European Communities) (1987f) Report on the Follow-up to the Council Resolution of 22 December 1986 on an Action Programme on Employment Growth, COM (87) 474 final, Brussels: CEC.

ELISE (1987a) Long-Term Unemployment: The Way Out?, Brussels: ELISE.

ELISE (1987b) Twelve European Regions Under the Microscope: The Dynamics of Local Development, Brussels: European Service Network.

Employment Gazette (1986) 'Action for jobs in Europe', 94 (10), November.

Harrison, J. (1986) 'A network for the future', Transition, November.

Irwin, P. (1987) 'European policies to help long-term unemployed people', Employment Gazette, November.

Johnstone, D. (1986) Programme of Research and Actions on the Development of the Labour Market: The Role of Local Authorities in Promoting Local Employment Initiatives, Luxemburg, Office for Official Publications of the European Communities.

Mawson, T. (1986) 'The training of long-term unemployed adults', Social Europe, 2/86, 81-3.

Morgan, S. (1986) 'Seminar on long-term unemployment', Social Europe, 2/86, 84-5.

Silletti, D. (1987) 'Specific measures taken under Community policy relating to the employment market: exchanges of officials from the national employment services', Social Europe, 1/87, 13-17.

Chapter Four

# LONG-TERM UNEMPLOYMENT: THE ROLE OF THE EUROPEAN SOCIAL FUND

Stephen Barber

The European Social Fund has traditionally played a key role in the fight against unemployment within Europe. As unemployment has risen during the 1970s and the 1980s, so people have come to look closely at what exactly the Social Fund is. This chapter takes a close look at the European Social Fund, at how it is administered, and how it operates. It looks both at the present Fund and at what is likely to succeed it during 1989-90. Particular attention is paid to the part of the Fund aimed at helping the long-term unemployed. The chapter is written deliberately very much from a practical viewpoint, reflecting the author's experience in being at the centre of the United Kingdom's operation of the Fund.

## WHAT IS THE EUROPEAN SOCIAL FUND?

Judged from the many and varied questions that the Department of Employment's Social Fund Unit receives, it is clear that some have little idea of what the European Social Fund is, while others sometimes clearly see it for what it is not.

The European Social Fund goes back to the Treaty of Rome (1957) itself and Article 123 of the Treaty which records the Social Fund as being established to: 'improve employment opportunities for workers in the Common Market and to contribute thereby to raising the standard of living'. It has as its task: 'rendering the employment of workers easier and of increasing their geographical and occupational mobility within the Community'. If one reads the 'rules' that cover the present Social Fund that began in 1984, one finds that in specific terms the Fund is very much about two things: vocational training and employment-creation schemes. In general, if the measure or programme fits neither of these labels, then the Social Fund is unlikely to be of help. It certainly has no direct link with 'social' schemes in the more general sense understood by many in Britain. The name derives from the close links that the Fund has with Ministers of Social Affairs - the equivalent of Britain's Secretary of State for Employment - within other member states. As will be shown later in the chapter, the Fund operates through a series of Council Decisions, Council Regulations, and Commission Decisions, all topped up with a degree of interpretation by the European Commission's civil servants working within its European Social Fund Directorate in Brussels. The life of a particular Fund is traditionally five years - although, because the Social Fund and its sister Regional and Agricultural Guidance Funds are currently in the midst of a major review, the present Fund will very probably run into 1989.

## THE PRESENT FUND

Targeted principally upon vocational training and employment-creation programmes, the current Fund nevertheless has a further series of priorities designed to focus its assistance on those groups deemed to be most in need.

**Those under 25 years old** Seventy-five per cent of the Social Fund allocations each year are channelled towards programmes covering the under-25s.

**Those who are unemployed or are threatened with unemployment, in particular the long-term unemployed** Under the Social Fund the long-term unemployed are those out of work for over twelve months.

**Women wishing to return to work** Although the whole of the Fund is open to women, there is a specific priority for schemes for women returning to work in occupations in which they are under-represented. But we know from experience that Commission officials look very closely at the occupations for which the particular women are being trained. They do not, for example, accept the argument that as women are under-represented in, say, management, the Social Fund should therefore be able to support programmes in this area. The Commission's Services' (civil service) response is that women are not under-represented in, for example, the management of florist shops or hair salons; it is the particular type of management in which women are under-represented that is relevant.

**Handicapped people capable of working in the open labour market** which, by definition, excludes a large number of disabled people.

**Migrant workers who move or have moved within the Community together with members of their families** But migrant workers exclude a large number of the ethnic minorities as they will have been in the United Kingdom for more than three years.

Finally, up to 5 per cent of the Fund is reserved for operations of an innovatory nature representing 'a potential base for future Fund assistance in the framework of the labour market policies of the Member States. These should test new approaches to content, methods or organizations of operations eligible for Fund assistance'.

In addition to specific groups that the European Commission sees as having a priority, the Social Fund also operates a system of geographical priority. There are three categories of priority: areas of absolute priority; priority; and others. The whole Fund is open to absolute-priority areas, less to priority areas, and significantly less to the others. The absolute-priority areas are decided at the beginning of the Fund and last for the life of the Fund (usually five years). The parts of the Community that currently attract this status are Greece, Portugal, Ireland, Mezzogiorno area of Italy, Northern Ireland, French Overseas Departments, and South and West Spain.

The priority areas are decided by the European Commission which compiles a league-table at United Kingdom county level and its equivalents throughout Europe,

**Table 4.1** The United Kingdom's priority areas (1989)

---

**Absolute priority**

Northern Ireland

**Priority**

**Counties/local-authority areas**
Central Scotland, Cleveland, Clwyd, Cornwall, Derbyshire, Dumfries and Galloway, Durham, Fife, Greater Manchester, Gwent, Gwynedd, Highlands, Humberside, Isle of Wight, Lothian, Merseyside, Mid Glamorgan, Northumberland, Nottinghamshire, South Glamorgan, South Yorkshire, Staffordshire, Strathclyde, Tayside, Tyne and Wear, West Glamorgan, West Midlands.

**Travel-to-work areas**
Workington and Copeland district in County of Cumbria
Llanelli in Dyfed
The Districts of Dover and Thanet in Kent
Accrington, Blackburn, Burnley, Lancaster, Nelson, Rossendale and Blackpool in Lancashire
Coalville in Leicestershire
Parts of Grimsby, of Hull and of Scunthorpe in Lincolnshire
Corby in Northamptonshire
North Warwickshire and Nuneaton and Bedworth Boroughs in Warwickshire
Bradford, Castleford, Dewsbury, Halifax, Huddersfield, Leeds, Keighley, Todmorden and Wakefield in West Yorkshire

---

based on levels of unemployment and Gross Domestic Product (GDP). A cut-off point is decided upon by the Commission in line with the proportion of the Community's working population that they feel the Fund can afford to support in the coming year. The list of priority areas is revised each year, based on fresh figures. As one of the main ingredients of the index is unemployment, the priority areas will, in the main, reflect those areas where long-term and youth unemployment is highest. As the table is revised each year, it can act as a reliable indicator of need. In addition to the unemployment/GDP element an area at a level below a county (in the UK) may attract priority status because it is assisted by the non-quota section of the European Regional Development Fund (ERDF) or under Article 56 of the European Coal and Steel Community Treaty. The full list of priority areas for 1989 for the United Kingdom is indicated in table 4.1.

## THE SOCIAL FUND GUIDELINES

Reference has already been made to the formal rules of the Fund established at the beginning of each five-year period. These rules are supplemented each year by a set of 'guidelines' or criteria setting out which types of operation the Fund will support in each of the geographical categories. Thus the guidelines that will apply for 1989 applications fall into five sections: general; operations for those aged under 25 years; those aged over 25 years; those with no age limitation; and innovatory projects.

Such is the increasing popularity of the European Social Fund that over recent years the Fund has been heavily oversubscribed. The European Commission's main response to this supply-demand imbalance has been two-fold: to reduce the number of areas attracting priority status; and to increase the number of guidelines applicable only to the priority or absolute priority areas. These moves have been only partly successful, and the European Commission has adopted a further measure known as 'linear reduction'. This measure has hit the adult part of the Fund in particular. It is a crude, yet effective, method which simply cuts by the required amount the size of each successful application to the Fund.

## SOCIAL FUND AND LONG-TERM UNEMPLOYMENT

Depending upon whether those on a particular programme or scheme, which is the subject of a European Social Fund application, live in an absolute-priority, a priority, or an 'other' area, and are long-term unemployed, they can probably be covered by any guideline. Applications under most of the guidelines will, therefore, include some long-term unemployed people. But there is one guideline for 1989

**Table 4.2** Number of UK successful applications under guideline 4.5 (1988)

| Organization type | Number |
| --- | --- |
| Companies limited by guarantee | 15 |
| Registered charities | 60 |
| Private companies | 2 |
| Local authorities | 257 |
| Educational establishments (including colleges of further and higher education and polytechnics) | 33 |
| Universities | 5 |
| Training Commission applications | 8 |
| Other government department schemes | 3 |
| Total | 383 |

applications specifically for the long-term unemployed, guideline 4.5. 'Vocational training of the long-term unemployed geared to their needs and including motivation and guidance and offering substantial prospects of employment'. By looking at applications under this guideline one can see, within the United Kingdom, which type of organization for 1988 mounted vocational training for the long-term unemployed which successfully attracted Social Fund support (see table 4.2).

An analysis of applications under this guideline by EC member states is shown in table 4.3.

It is perhaps worthwhile having a close look at the actual guideline and how it is interpreted by the European Commission's Services. To attract European Social Fund support, the scheme must offer vocational training which 'equips trainees with the skills required for one or more specific types of job'. It must have a minimum duration of 200 hours, including 40 hours devoted to training broadly related to the new technologies. The training must 'be geared to their needs', which the Commission Services define as requiring it to be wholly or mainly for the long-term unemployed. To run a scheme which trains the long-term unemployed alongside other unemployed could,

**Table 4.3** Applications by EC member states under guideline 4.5 (1988)

| Member state | European Social Fund applications submitted (1988) | Applications submitted under guideline 4.5 (1988) | % |
|---|---|---|---|
| Belgium | 448 | 89 | 20.0 |
| Denmark | 109 | 4 | 4.0 |
| France | 981 | 69 | 7.0 |
| FR Germany | 493 | 21 | 4.5 |
| Greece | 821 | 95 | 11.5 |
| Ireland | 120 | 12 | 10.0 |
| Italy | 1,540 | 92 | 6.0 |
| Luxemburg | 13 | 0 | – |
| Netherlands | 725 | 129 | 18.0 |
| Portugal | 1,439 | 53 | 3.7 |
| Spain | 679 | 58 | 8.5 |
| United Kingdom | 2,909 | 428 | 14.7 |
| Totals | 10,277 | 1,050 | 10.2 |

Note Not all of the applications submitted will have been successful.

therefore, reduce the chances of the scheme successfully attracting European Social Fund support. The training must offer substantial prospects of employment. If less than 50 per cent of those trained are likely to get jobs at the end of the course, the application is unlikely to be supported. Finally, and perhaps of greatest significance, 'the long-term unemployed guideline' is only available in absolute-priority and priority areas. The long-term unemployed in areas such as inner London, therefore, are excluded - and this at a time when the European Commission is agreed upon the priority that is to be accorded to long-term unemployment. Some observers have commented upon the apparent contradiction in this.

The Training Commission (the former Manpower Services Commission) has been successful in being allocated

£24 million for the Employment Training programme, scheduled to start in September 1988. This was seen as an important new contribution and dimension to a major programme aimed at tackling long-term unemployment.

## THE FUTURE OF THE SOCIAL FUND

The present five-year European Social Fund was due to end in December 1988. In July 1987 the Single European Act was finally ratified by all twelve member states. Article 130D of the Treaty provided for a review of the Structural Funds, to clarify and rationalize their tasks. This review is now well under way against the background of the decision of the European Council in February 1988 to double the Structural Funds between 1987 and 1993. In June the European Council agreed the broad framework that would form the basis for the new Funds and gave some indication of what their shape would be. This 'Framework Regulation' set out five priority objectives for the Funds:

promoting the development and structural adjustment of the less-developed regions;
converting the regions, frontier regions, or parts of regions (including employment areas and urban communities) seriously affected by industrial decline;
combating long-term unemployment;
facilitating the occupational integration of young people;
with a view to reform of the Common Agricultural Policy, speeding up the adjustment of agricultural structures, and promoting the development of rural areas.

Although the European Social Fund will contribute towards each of the objectives, its main tasks will be concerned with the long-term unemployed and young people who have completed their period of compulsory full-time education. Emphasis in the Framework Regulation is placed upon the need for a partnership between the European Community and member states, and Community-funded operations are seen as being complementary to national measures. Close consultation between all concerned in pursuit of a common goal is seen as essential.

At present the Social Fund assistance is allocated

mainly to a large number of sometimes small individual project applications (the United Kingdom alone submitted almost 3,000 in 1988). This process will change under the new Fund when the emphasis will be on longer programmes of up to three years' duration. There will also be closer monitoring and evaluation of operations to gauge their effectiveness. Assistance will be concentrated upon less developed regions. Within the United Kingdom only Northern Ireland will qualify for this status. The United Kingdom is also expected to benefit under Objective 2, covering areas of industrial decline based upon unemployment, the share of industrial employment, and its rate of fall. At the time of writing the list of Objective 2 areas is not known.

The broad framework for the new Funds has been agreed and will be followed by more detailed implementing regulations which, according to the European Commission's indicative timetable, will be the subject of negotiations during the rest of 1988, with their adoption scheduled by December 1988. As far as the European Social Fund is concerned, however, applications for 1989 are already being put together, so the new Fund is unlikely to start until 1990. The more detailed transitional arrangements will emerge over the coming months.

**Chapter Five**

## POVERTY, UNEMPLOYMENT, AND THE EUROPEAN COMMUNITY: LESSONS FROM EXPERIENCE

John Benington

Long-term unemployment is now firmly on the agenda of the European Commission and Parliament. With over 16 million people officially unemployed in the European Community (half of them out of work for more than one year, and about a third unemployed for more than two years) it is perhaps not surprising that long-term unemployment has been confirmed as a priority objective for the reformed structural funds, and for the run-up to the completion of the single European market by the end of 1992. However, the close connection between long-term unemployment and poverty has not yet been so clearly established at the European level. It is also not so widely known that the European Commission has sponsored two European action-research Programmes to Combat Poverty, which have highlighted long-term unemployment as one of the key factors in the recent growth in poverty in the EC member states.

Although there has been continuing debate about how best to define poverty, there is no disputing the dramatic increase over the past decade. Using a definition of 50 per cent of the average household income in each respective state, a recent study estimated that the number of people in poverty in the twelve EC countries had risen from 38.6 million around 1975, to almost 40 million in 1980, and then to around 44 million in 1985. (1) It is estimated that the 1988 figure is at least 45 million.

Changes in the composition of the poor are as significant as the rise in numbers. A recent series of studies of the 'new poverty' for the European Commission showed that there have been two main changes in the composition of the poor in the member states - a substantial decline in the proportion who are elderly; and a sharp increase in the proportion who are unemployed, or in very low-paid or insecure employment. (2)

The aim of this chapter is to describe and analyse the activities and findings of the 2nd European Programme to Combat Poverty in relation to long-term unemployment, and to discuss their implications for European economic and social strategies.

## THE EUROPEAN PROGRAMME TO COMBAT POVERTY

The 1st European Programme to Combat Poverty ran from 1975 to 1980 and consisted of 49 local action-research projects in eight member states. The 2nd European Programme was to run from 1985 to the end of 1989 and included 91 action-research projects across the twelve member states. Its budget was small (20 million ECU in total, which works out at less than 0.20 ECU per head of the European population in poverty). However, its aims were big:

> To combat poverty more effectively and carry out positive measures to help the under-privileged, and identify the best means of attacking the causes of poverty and alleviating its effects in the Community (Article 1.1 of the European Council decision of December 1984).

The definition of poverty used by the Council of Ministers was equally bold:

> For the purposes of this Decision 'the poor' shall be taken to mean persons, families and groups of persons whose resources (material, cultural and social) are so limited as to exclude them from the minimum acceptable way of life in the Member States in which they live (Article 1.2).

Nevertheless, the European Programme to Combat Poverty has been greeted with mixed feelings in some quarters. On the one hand, the fact that 91 local projects have the opportunity to take part in a four-year European action-research programme is welcomed (particularly at a

time when poverty is increasing and intensifying, so rapidly).
On the other hand, there is considerable scepticism among
some of those in the 'poverty lobby' about the usefulness of
yet another relatively small-scale, short-term, locally based
programme, coming as it does after a succession of similar
low-cost initiatives to investigate poverty and deprivation
over the past 17 years. Many critics argue that there have
been more than enough investigations, more than enough
pilot projects, and that the need now is for changes in
mainstream government policies to act on the findings and
recommendations from previous anti-poverty programmes.
These criticisms have some force, particularly when they
come from the residents of local communities which have
been the focus of earlier anti-poverty measures in the UK,
Ireland, or the other member states.

At the same time the 2nd European Programme has a
number of distinctive features that mark it off from its
predecessors:

> It is on a European-wide scale, with 91 local projects
> drawn from all twelve member states of the European
> Community. This allows the possibility of a very
> 'textured' analysis of what it means to be poor in
> Europe in the 1980s - in urban and in rural situations, in
> industrialized regions as well as in the less developed
> periphery.

> The projects have been selected in relation to cross-
> national themes, each concerned with a particular
> target group or issue rather than with poverty in
> general terms.

| | |
|---|---:|
| the long-term unemployed | 11 projects |
| the young unemployed | 9 projects |
| the elderly | 12 projects |
| single-parent families | 9 projects |
| second-generation migrants, refugees and returning migrants | 12 projects |
| 'marginal' and homeless people | 11 projects |
| integrated action in urban areas | 14 projects |
| integrated action in rural areas | 13 projects |

(It is notable that several key groups have been omitted
from this categorization, e.g. women and black and
ethnic minorities).

The European Commission has delegated some tasks to an independent Animation and Dissemination Service with the following functions:

to promote collaboration and exchange of experience among the projects in the Programme. This is undertaken by a team of eight theme co-ordinators, working part-time on a cross-national basis, under the oversight of an overall co-ordinator or co-ordinators;

to monitor the projects and to evaluate the significance of their activities. This is undertaken by a team of ten evaluators, working part-time on a national basis, and led by a research team at the Centre for Analysis of Social Policy at Bath University;

to promote dissemination of the results of the Programme throughout the Community. This is carried out by the Institut für Sozialforschung und Gesellschaftspolitik (ISG) in Cologne, West Germany.

An unusually high level of priority has been given by the European Commission (and the co-partners and the projects) to monitoring and evaluation of 'the progress and the productiveness' of the projects with reference both to their individual programmes of work, and to the overall objectives of the Programme.

As shown above, 20 of the 91 projects are specifically focused on long-term unemployment or youth unemployment. However, a much wider range of the local projects (concerned, for example, with rural and urban development, single parents, the elderly, migrants, and the homeless) have also found unemployment to be a key factor in the poverty of their target group or area. They have, accordingly, identified the need for specific employment measures (e.g. social support, skill training, job creation, and economic development) as part of their local strategies to combat poverty.

It is obviously important for the European Commission and the member states to take account of this small-scale but substantial body of European-wide action-research into poverty, as part of its wider commitment to combating long-term unemployment and integrating young people back into the labour market. The first part of this chapter will describe and categorize the range of types of unemployment by which the projects have been confronted at the grassroots level. It will then briefly describe and categorize the range of action strategies that they are developing. Third, the chapter will provide an analysis of the main causes of the growth in unemployment in these areas and the ways in which this growth leads on to poverty. Finally, the chapter will suggest some of the implications of the projects' work for European and national economic and social strategies.

## FOUR TYPES OF UNEMPLOYMENT

The links between unemployment and poverty are not self-evident. However, analysis of the action-research by the local projects has identified four main types of employment problem which contribute to poverty within the project areas.

### Cyclical unemployment

Several projects are working with groups of people who have been thrown into unemployment and poverty mainly as a result of cyclical changes in the economy. The contraction of both private and public investment which took place in many West European countries during the late 1970s and early 1980s has led to sharp increases in unemployment among people who have skills that are still likely to be in demand as and when the economy picks up (e.g. construction workers). Projects have found that, although people in this category were in stable jobs and situations until recently, unemployment can push them down the vicious circle into poverty and dependency very rapidly. The loss of a job leads not only to loss of income, debt, and other financial pressures, but also to loss of status, confidence, and social relationships. The slippery slope from stable employment via unemployment into poverty and demoralization can be a very sudden one. Similar problems affect young people who have left school with qualifications, but then suffer long periods of unemployment. The consequences get more serious and the costs of reinstatement into full-time employment get greater, the longer the term of unemployment suffered. Projects' experience in combating these trends has far-reaching implications for EC and national employment policies.

## Structural unemployment

A large number of projects are dealing with unemployment arising from the relocation or restructuring of traditional industries, technologies, and skills, or the introduction of new technologies. This kind of unemployment tends to be concentrated in those European regions and localities which grew up to provide labour and other resources for the key industries of the previous industrial revolution (e.g. coal, steel, shipbuilding, textiles, metal manufacture). Similar processes are at work, however, in some rural areas suffering from the European restructuring of the agriculture industry. Many of the anti-poverty projects are set in such areas of concentrated unemployment and economic and social decline. They are finding that unemployment in such areas is particularly intense, because factories and workplaces have closed or run down, the environment has deteriorated, and large numbers of workers have been made redundant, often at very short notice. In this case the general problems associated with unemployment and poverty are compounded by the demoralization arising from people's feeling that their skills have become obsolete and worthless.

## Underemployment

Some projects are finding that poverty in their areas arises even when people are in paid jobs. This poverty arises because the decomposition and segmentation of the labour market and the erosion of employment rights mean that a growing proportion of jobs (particularly in the service sector) are part-time, temporary, casual and/or low paid. In several of the project areas the unemployed are having to accept part-time, casual, temporary, or insecure employment on very low wages as the only alternative to long-term unemployment. In this case poverty is related to underemployment rather than complete unemployment.

## Hidden unemployment

In addition to the above forms of unemployment and poverty, projects are confirming evidence of other research which shows a layer of hidden unemployment not recorded in the official statistics. This unemployment includes groups like older workers, young people on temporary employment or training schemes, women (particularly single parents), migrants, and black and ethnic minorities who wish to work but are discouraged or even excluded from paid employment by:

financial disincentives (e.g. poverty traps within the social security system);
social discrimination (e.g. racism, sexism, ageism);
the lack of the support systems necessary to make employment outside the home possible (e.g. child-care facilities for single parents).

## A poverty underclass?

A further category of unemployment and poverty in the project areas is those people who might be expected to have employment difficulties, even in a situation of full employment. There is considerable debate and disagreement as to whether it is accurate or helpful to describe this group (in the American terminology) as an 'underclass'. However, several of the projects are undoubtedly working with people in poverty who want (and benefit from) work experience and employment opportunities, but who need very long-term support, in a sheltered work environment, before being able to compete in the open labour market for full-time paid jobs.

## PROJECT STRATEGIES FOR TACKLING UNEMPLOYMENT AND POVERTY

It is clear from the above that the projects are dealing with a very wide spectrum of employment problems. These range from the 'new poor', who until recently were in full-time skilled jobs, who have been pushed into unemployment and poverty directly as a result of the economic recession or industrial and technological restructuring, and who would be lifted out of poverty and dependence on welfare benefits simply by a targeted programme of job creation and (re)training. At the other end of the spectrum are those facing more complex pressures and difficulties, disincentives and discriminations within the labour market. The solution to their poverty and employment problems would also require changes in social-security policy to remove poverty traps; legislative measures to combat discrimination within the labour market; and social-support systems (e.g. adequate child-care provision). There are also some who would require long-term support within a sheltered-employment environment.

However, one important conclusion arising from the work of the projects is that people do not necessarily remain at fixed points on this scale. Those only recently in full-time skilled employment can slide very rapidly down the slippery slope into poverty and despair. Equally, those at the bottom of the ladder of demoralization and dependence can move out of poverty and back into employment and self-sufficiency if helped to regain confidence and skills by an integrated programme of social support, skill training, and tailor-made job creation. The link between employment, unemployment, and poverty is not a steady continuous one but a series of sharp steps up and down (more like the game of snakes and ladders). This process is illustrated in figure 5.1.

Project strategies to tackle unemployment and poverty have been geared to helping groups resist the pressures pushing them down the slippery slope from unemployment and poverty into demoralization and despair; and, wherever possible, to hold on to (or to rebuild) their confidence and skills, through group support and skill training, to regain employment.

Six main strategies can be identified amongst the projects.

## Group support, solidarity, and counselling

Many projects aim to overcome the social isolation, guilt, and demoralization associated with all forms of unemployment in society by group activities which restore a sense of solidarity and belonging, and therefore of confidence and worth. For example, Project 14 in Dortmund, West Germany, and Project 49 in Whitley Bay, United

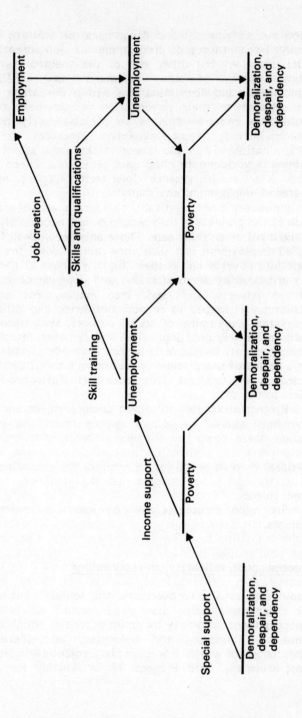

**Figure 5.1** The link between employment, unemployment and poverty

Kingdom, provide individual and group counselling on questions connected with the personal, economic, legal, and social situation of the long-term unemployed. Project 50 in Wolverhampton, United Kingdom, helps to set up autonomous organizations of the unemployed who maintain their morale through self-help group activity, and campaigning on welfare rights and other issues.

## Training in work and social skills

Several of the projects are experimenting with new forms of training in work and social skills. These projects are often (as in Project 61 in Belfast) tailor-made to meet the precise employment histories and training needs of the local employed as well as skill shortages and employment opportunities within that particular locality. In other cases they are targeted on groups who often receive little priority under existing provisions. For example, Project 3 at Charleroi, Belgium, provides a series of apprenticeships for the young unemployed with few formal educational qualifications. Project 42 in Milan, Italy, provides vocational training courses for the long-term unemployed. Project 26 in Evry, France, aims to provide job training for single parents. Project 4, Service Sociale in Belgium, targets training and other forms of support at a refugee population which is socially and culturally integrated, but where unemployment puts people at risk of slipping into an underclass position.

## Provision of work experience in socially useful employment

Some projects provide sheltered or simulated work experience (or temporary work experience on an employer's premises but without full pay or permanent contract) as part of the transition from poverty and unemployment into full-time paid employment. Several of the Belgian and French projects include this type of experience within their strategy, e.g. Project 3 at Charleroi targeted at young people; Project 5 at Viboso, working with homeless people and other marginalized groups; Project 45, Inter-Actions Faubourg, focused on the young unemployed; and Project 46, Femmes en Détresse, targeted at women with children. Other projects working in this way include Project 22 in Lens, Project 25 at Plantaurel, and Project 27 at Jazeneul,

93

France, all focused on the homeless.

## Job creation

Although operating with very limited resources and powers, several of the projects are collaborating with other agencies with the long-term aim of creating new jobs. These projects are based on the awareness that efforts to help the unemployed (re)gain confidence and skills can be completely wasted if there are no jobs available for them at the end. They can then all too quickly and easily slip back down into poverty, demoralization, and even homelessness.

The Belfast and Derry projects in Northern Ireland (61a and 61b) are preparing several job-creation projects designed to meet both unmet local needs and the skills of the unemployed. Some of these may be workers' co-operatives or community businesses. Several of the rural development projects are carrying out feasibility studies for major job-creation projects to help diversify employment opportunities within the local economy. Projects 32 at Louisburgh, 33 at Connemara, and 34 at Inishowen in the Republic of Ireland are developing proposals for projects in forestry, mariculture, and tourism. Similarly Project 60, working in four rural areas of Northern Ireland, is testing the feasibility of alternative farming, tourism, and cultural projects.

## Economic development

For some of the projects (particularly those concerned with the themes of integrated rural and integrated urban development) skill training and job creation are set within the wider context of plans and proposals for the economic development of their whole area. This is particularly true of the Greek and Italian projects and some of those in the Republic of Ireland and Northern Ireland. (See also the chapter by Kennedy on Bradford).

Project 17 in Ikaria and 21 in Crete, Greece, are both developing integrated plans for tackling the poverty and underdevelopment of their region through new investment in locally determined co-operative projects (fishing, agriculture, dairy products, weaving, etc). Similarly Project 37 in Palermo, Sicily, Project 38 in Naples, and Project 39 in

Friuli, North East Italy, are all concerned with generating jobs to match the skills of the unemployed and the needs of their areas through economic development plans involving forestry co-operatives, environmental conservation, agricultural development, tourism, and artistic and cultural activities.

## Campaigning

Most of the projects realize that they themselves do not have the resources or the powers to combat the problems of poverty and (un)employment which they confront in their project areas. Accordingly, many of them complement their local action-research programmes with campaigning activity to bring the issues to the attention of regional, national, and European policy-makers. Such campaigns to raise consciousness of problems or to change policy can also have the effect of mobilizing the poor and the unemployed to higher levels of activity, participation, and autonomy. Project 30 in Darndale, Dublin, is run by the unemployed themselves and plays a prominent role in representing the unemployed in campaigns and in the media. The Belfast and Derry Unemployed Workers Centres (Project 61) have joined with trade-union and other community organizations in several campaigns concerned with the welfare rights of the unemployed, and for alternative economic strategies for their areas. The Wolverhampton project (number 50) has also campaigned for specific changes in social-security procedures, which hamper the involvement of unemployed people in cross-national meetings and exchanges.

## AN OUTLINE ANALYSIS OF THE CAUSES OF UNEMPLOYMENT AND POVERTY IN THE PROJECT AREAS

It is important to analyse the links between unemployment and poverty a little more closely in terms both of communities and of individuals. Although this analysis is undertaken by reference to the situation in the United Kingdom, it is also relevant to other northern industrialized regions of Western Europe and the Central European core.

In order to understand the changing context and patterns of poverty in the local project areas, it is necessary to understand the rise in unemployment and low pay, and their extension to wide sections of the population. Both are related partly to European industrial restructuring and partly to economic and social policies in the EC member states.

## Poverty resulting from restructuring of industry and employment

OECD standardized unemployment figures illustrate a sharp rise in unemployment in all the major industrialized European countries (table 5.1).

**Table 5.1** Rise in unemployment since 1979

|         | February 1987 % | Rise in percentage points since 1979 |
|---------|-----------------|--------------------------------------|
|         |                 | %                                    |
| UK      | 11.1            | 6.1                                  |
| France  | 11.0            | 5.1                                  |
| Italy   | 10.7            | 3.1                                  |
| Germany | 8.0             | 4.8                                  |
| Canada  | 9.6             | 2.2                                  |
| USA     | 6.6             | 0.8                                  |
| Japan   | 3.0             | 0.9                                  |

These increases cannot be explained purely in terms of an increase in the supply of labour (although in the UK there has been some rise in the 'supply' of school leavers and women). The main explanation lies in the drastic 'deindustrialization' which has been taking place. Again, the trend began in the 1960s but accelerated very sharply after 1979. Employment in manufacturing in the United Kingdom fell by almost 1.5 million between 1966 and 1979, and by a further 2 million from 7.1 million to 5.1 million (a fall of 28 per cent) between June 1979 and March 1987.

Much of the loss in manufacturing jobs has taken place in traditional 'smokestack' industries - coal, steel, shipbuilding, metal-working, motor vehicles and textiles. Many of these industries are experiencing classic problems of overproduction, which then leads through a process of cut-throat competition, merger, takeover, or closure, to greater concentration and centralization of control within that particular sector. In many cases this restructuring is taking place on a European or worldwide basis, and concentration and centralization is leading to closure or run-down of 'peripheral' sites in the declining regions. European Community policies have in some cases (e.g. steel, textiles) assisted or accelerated the process of restructuring to deal with overproduction, and provide subsidy schemes to encourage reductions in capacity in member states. In other cases (e.g. the motor and components industry) transnational firms have shifted investment and production to overseas subsidiaries and run down their workforces in particular branch plants.

The implication of this process for local project areas is that many of these traditional industries have tended to be concentrated spatially in regions and localities which grew up to provide them with the labour they needed, at earlier stages in their life cycle. Restructuring of these industries has, therefore, had a particularly sharp impact on specific communities. The impact has also tended to be sudden and unpredictable.

Restructuring has rarely taken place in a steady way, but through crises like plant closures, relocations, or mass redundancies. The rise in unemployment and poverty in several of the local project areas is mainly attributable to the haemorrhage of job loss resulting from contraction of these traditional industries:

> coal and steel in the South Wales project areas, and in the eastern Ruhr region where the Dortmund project is based;
> metal working and motor vehicles in the West Midlands and Wolverhampton;
> shipbuilding in the North East (Whitley Bay) and Northern Ireland (Belfast);
> textiles in Bielefeld in West Germany;
> longer-term changes have led to the decline or marginalization of agriculture in regions like Lancashire (ERA) and Northern Ireland (Rural Action Project), with slow deterioration of their economic base.

### Job loss and unemployment spread to the service sectors

The reference to agriculture at the end of the previous paragraph is a reminder that job loss is not restricted to the manufacturing sector. It has also affected agriculture (though in this case the consequences of European overproduction are temporarily cushioned by the Common Agricultural Policy). Moreover, it has now extended to the service sector. The experience of the United Kingdom may be indicative of wider trends in Europe.

Although the drastic loss of manufacturing jobs in the United Kingdom was compensated to some extent until the mid-1970s by an increase in service-sector employment, this growth too levelled off around 1975 and had actually begun to go into reverse by the beginning of the 1980s. About a quarter of a million service-sector jobs disappeared in 1980 alone. Although some areas of the service sector appear to be growing again (e.g. the financial-services sector), even these areas are liable to contract as a result of the introduction of new technologies. Furthermore, many of the new jobs in the service sector are part time and insecure.

The overall situation in the United Kingdom and in the regions and localities in which the projects are situated is that total employment has contracted. The number of employees in the United Kingdom has dropped by 1.4 million since 1979, from 22.6 million in June 1979 to 21.2 million in December 1986, a fall of 6 per cent. Unemployment and poverty in the United Kingdom and in the project areas have risen mainly as a result of the job loss arising from the restructuring of industry and employment, and the failure to generate new industries and new jobs in the right areas and in sufficient numbers to replace them.

### The impact of technological change on poverty and unemployment

The restructuring of traditional manufacturing industries has been one of the main reasons for the drastic rise in unemployment and poverty in the European Community over the past decade. Research suggests that the European Community needs to prepare for a further worsening and intensification of the pattern of poverty as the next wave of technological change hits the service sector, with potentially as devastating consequences for jobs and skills in the service sector in the 1990s as deindustrialization had on the manufacturing sector in the 1970s and 1980s.

As in previous epochs of technological change (the stone age, the bronze age, the steel age), the new tools that are now being developed open up the opportunity for fundamental changes in society and culture - not only in patterns of employment but also in the patterns of communication, human settlement, domestic life, and leisure. While there may be differing views about the precise impact of the changes over the next decade, the new generation of technologies undoubtedly represents more than just another incremental change. It is more like a quantum leap into a new era. This leap is a consequence of the linking up of three technologies: computing (with the ability to process vast amounts of information at enormous speed); micro-electronics (with its extraordinary possibilities for miniaturization and dramatic reductions in costs); and telecommunications (which can open up remote access to, and interaction with, computing and micro-electronics across international networks). These three technologies, when linked together, open up entirely new possibilities for the automation of mental functions as well as manual processes; for office, design, and managerial work as well as shopfloor work; for small-batch production as well as mass production; and for the whole business-management and information system (including planning and co-ordination) as well as individual parts of the production process.

The impact of computer technology on jobs and skills in manufacturing industry is well advanced and well documented. Much less is understood and documented about the service sector. However, it seems likely that this sector will be the focus of the technological revolution over the next decade, with the widespread replacement of paper-based information systems by computer- and tele-communication-based systems. Some detailed studies by employers, trade unions, and academic bodies estimate the impact on service-sector employment to range from 20 per cent to 40 per cent job loss by 1995. Siemens of West Germany suggests that 40 per cent of present office work could be carried out by computer systems by 1990. The International Federation of Commercial, Professional, Clerical and Technical Employees (FIET) has estimated that by 1990 between 20 per cent and 25 per cent of the current 15 million office jobs in the EC will have been affected by the new technologies; 30 per cent of banking and 20 per cent

99

of retail jobs could also be lost.

The implications of these trends can begin to be seen more sharply when some of the possible patterns in the distribution of unemployment and employment, and in the skills required for the future, are also highlighted:

> an increase in the proportion of the population whose skills are at risk from redundancy, or who require retraining in order to remain employable. This increase is already evident not only in manufacturing but also in the commercial and service sectors - banking, insurance, office, retail, etc. - and in white-collar and managerial occupations;
>
> an increase in the pattern of home-based working as large enterprises exploit the opportunities provided by computer, telefax, cable, and other new communication systems to decentralize and/or to sub-contract parts of their operations;
>
> a deepening division between, on the one hand, the declining regions where high rates of unemployment will persist as deindustrialization of primary and traditional metal-based industries continues and, on the other hand, the growth regions where a virtuous circle of prosperity may be kept in motion by the high-tech and financial sectors and by a new generation of design-led, small-batch production, and distribution industries based on computer-flexible specialization; (3)
>
> a particularly high incidence of unemployment among the over-55s;
>
> growth in the number and proportion of the very long-term unemployed (over two years);
>
> continued heavy concentrations of the unemployed in the inner-city areas, and particularly within the black and ethnic minority communities (where unemployment is already frequently about 50 per cent).

## ANTI-POVERTY POLICIES AND DEBATES

The profound changes in the patterns of unemployment and poverty in Europe as a result of the restructuring of industry, technology, skills, and jobs have been matched by equally fundamental changes in government economic and social policies in many of the EC member states. More important than the individual policy changes, however, is the revolution in underlying philosophy introduced by the New Right since the late 1970s. This development marks a decisive watershed in the history of the Western welfare state and a challenge to the whole post-war Keynesian consensus in Europe.

Part of the New Right revolution has been the attempt to cut back central and local state expenditure on the grounds that its growth has 'crowded out' private investment and has been too heavily parasitic on the wealth created in the private sector. However, the challenge to the welfare state has been ideological as well as economic. Monetarists have argued the need to 'roll-back' the frontiers of central and local government, and to privatize and to contract out as many public services and industries as possible, as part of a radical challenge to the whole concept of a collectivist welfare state.

The New Right philosophy is one in which needs are to be met primarily through the private market (or through self-help and the nuclear family) rather than through the public sector. Their model for society is a market of consumers purchasing individual services, rather than a community of citizens making collective provision for their needs through the state. This philosophy involves a decisive shift away from the founding principles of the post-war welfare state - a system based on the assumption of full employment, plus national insurance to provide universal benefits for the additional needs arising at key periods of life (e.g. maternity, old age, death) and to insure against the main risks (e.g. sickness, unemployment, disability). The principle of universal services, provided as of right, through a communal insurance scheme (backed up by a safety net of supplementary benefits for special needs and emergencies), is being challenged by an alternative philosophy of individuals meeting their own needs through self-help and the market - private provision through private insurances. The aim is to reduce the state to a minimal welfare role, providing mainly for minorities who have not succeeded in paying for their own services through the market. State benefits are, therefore, designed to be kept to a means-tested minimum, in order not to provide any disincentive for people to aspire to meet their own needs again through the private market. Inevitably, therefore, an element of division and stigma is introduced into the social-security system in place of the original concept of universal entitlement. When the notion of rights based on national insurance is eroded, the distinction between deserving and undeserving poor is reintroduced.

The notion of differentials as a necessary spur for competition and enterprise is a crucial part of the New Right philosophy for restructuring British society. For them the welfare state has encouraged dependency and passivity and removed the incentives necessary for free enterprise. Within this model tax incentives to encourage and reward the rich are paralleled by disincentives (e.g. lower wages and benefits, more stringent means tests) to discourage the poor from lapsing into dependence on the state.

The argument has been that such measures are necessary because workers have priced themselves out of jobs. However, there is little evidence that the lowering of wages has led to the clearing of labour markets or to any measurable increase in the number of jobs, though it may have helped the restoration of profitability in certain sectors. The paradox within the whole policy is well expressed by J.K. Galbraith: 'It holds that ... the poor do not work because they have too much income; the rich do not work because they do not have enough income.' (4) Poverty and inequality therefore play a paradoxical role within New Right policy. On the one hand, the lowering of wages and state benefits are seen as a necessary spur to the creation of a more internationally competitive and flexible economy - the 'stick' that complements the 'carrot' of high incomes and low taxes for those who succeed. On the other hand, these policies have resulted in an increased burden of spending on social-security and supplementary benefits, thus thwarting their avowed intention to cut back public expenditure.

Unemployment and poverty are, accordingly, at the centre of a battle of fundamental ideas and values about society. The increasing polarization within European political and economic thinking, however, should not obscure the fact that there is widespread agreement that existing policies and programmes to tackle unemployment and poverty fail to match the changed scale and nature of the problem.

They were based, after all, on the assumption of full employment. The insurance 'safety net' has undoubtedly been strained to breaking point by having to deal with a high level of long-term unemployment and increasing poverty.

The financial costs of providing for an increased proportion of dependents (unemployed and elderly) do represent a 'fiscal crisis' when the economy is no longer growing (though the UK could have used oil revenues to overcome the fiscal crisis as Norway has done).

The provision of services by large centralized public bureaucracies has been accompanied in too many cases by considerable inefficiency and insensitivity (see the chapter by Young).

The real debate in Western Europe is not whether the welfare state needs reforming, but how. How to deal with poverty and unemployment are key issues within those debates. In fact, they have become in many ways archetypes or metaphors for fundamentally different philosophies about the economy and society as a whole. The underlying debate is not really about levels of public expenditure, but about the roles of the state and the private sector, and about collectivist or private-market methods of meeting human needs. Policies for dealing with unemployment and poverty are, therefore, no longer marginal social-welfare issues, but central to much wider questions about the direction for European society as a whole.

## THE IMPLICATIONS FOR NATIONAL AND EUROPEAN POLICY

Earlier sections of this chapter have referred to some of the projects' interim findings about unemployment and poverty. These findings can now be restated and brought together in a series of propositions, with implications for European and national policies.

### The new poverty

The new poverty confronting the European Community is quantitatively and qualitatively different from earlier patterns of poverty. It is quantitatively different because it now affects much larger sections of the population, in terms of numbers, age range, gender, skill distribution, and socio-economic class, ethnic background and regional profile. It is qualitatively different because whole new categories of people have been brought into poverty as a direct result of the restructuring of the economy and of industry in the EC member states with a high level of job loss in traditional 'smokestack' industries and the areas which grew up to support them.

## Multiple deprivation

The material consequence of this new poverty is that large numbers of people who would otherwise have been able to support themselves through their employment or through state insurance benefits are now thrown into multiple deprivation - suffering a vicious circle of unemployment, poor housing, run-down environment, inadequate health services, low-quality education, and inadequate transport facilities.

## Powerlessness and marginalization

These multiple material forms of poverty are often accompanied by a sense of powerlessness and marginalization. Although experienced as an individual psychological problem, powerlessness is obviously related to the structural position of the poor. Their exclusion from paid employment is a fundamental cause of their powerlessness and marginalization, as work still confers the main source of status and value in European society.

Loss of wages and low income exclude the poor from one of the major sources of power in European society under present conditions: the power to purchase goods and services in the private market. This factor reduces not only their power to purchase commodities but also their access to many communal situations such as sport and leisure facilities.

## Disinvestment and underdevelopment

The projects have been confronted not just by the poverty facing individuals and groups but also by the poverty of their areas. Several of the projects have analysed this situation in terms of the run-down or withdrawal of both private and public investment, often as part of a national or European-wide restructuring of key industries like agriculture, coal, steel, shipbuilding, motor vehicles, and textiles. Plant closure, relocation, or redundancy have led not only to long-term unemployment for individuals, but also to loss of business for local suppliers, sub-contractors, and shopkeepers, a loss of rate income for the local authority, a drop in spending power within the area, and a vicious circle of economic and social decline.

The policy implications of the above analysis of poverty and unemployment are far reaching. If poverty is no longer a social-welfare problem affecting small minorities who have slipped through the safety nets of the social services, but a symptom of a much deeper restructuring of the economy, industry, and the welfare state, then it follows that it cannot be tackled merely by modifications or innovations in local welfare services. It requires changes in economic and social policies at national and European levels.

## Existing programmes sometimes make the problem worse rather than better

Many of the existing forms of governmental support for the long-term unemployed (including some of those financed via the European Social Fund) fail to achieve their declared aims, and in some cases actually make the problem worse rather than better.

Financial assistance is often too low to allow even minimum subsistence-level living. This situation can lead to further problems of debt, ill-health, etc., which lead to pressures on public services.

It is often provided in ways that stigmatize and reinforce the loss of dignity experienced by the unemployed.

Social and welfare-support services often reinforce the unemployed person's sense of powerlessness and marginalization, rather than assist him or her in regaining confidence and autonomy.

Too many of the training schemes for the long-term unemployed fail to provide skills which are required within the local labour market, or fail in their job-placement arrangements. High percentages of the trainees thus slip back into unemployment and poverty at the end of training, rather than being helped on to employment. This outcome is demoralizing for those concerned and a waste of the public resources invested in the training programme.

Some job-creation programmes in areas of high unemployment fail because they do not relate to the indigenous technologies or the skills and potential of those looking for work, and thus do not provide jobs for those who need them most.

## The need for more cost-effective strategies

The projects have, therefore, growing evidence that existing policies and programmes within the EC member states are inadequate to cope with the magnitude and gravity of long-term unemployment, or the scale and intensity of the new poverty. Much government expenditure is being wasted on programmes that are not cost-effective because they fail to reduce or ameliorate the problem. In a sense they pour resources down a bottomless pit. The post-war welfare policies which were designed to provide a safety net for relatively small percentages of the population who fell into unemployment for relatively short periods of time are simply not coping with supporting the massive numbers of people now facing unemployment and poverty for long periods of time. Radical new strategies are required. Instead of aiming simply to provide remedial support, these strategies must be designed to lift the long-term unemployed out of poverty by getting them back into productive and socially useful employment.

## The opportunity for the European Commission to give a lead

The European Parliament and the European Commission have a particular opportunity and responsibility to help the EC member states to tackle this intransigent problem. The growth in long-term unemployment has a European dimension in the sense that it is at least partly caused by the European restructuring of traditional industries (e.g. agriculture, coal, steel, shipbuilding, textiles, metal working) and by changing European patterns of consumption (e.g. tourism and leisure, and its effects on the Mediterranean regions). Furthermore, many of the existing programmes of support, training, and job creation for the long-term unemployed within the EC member states are heavily subsidized by European Commission funds. The review and reform of the European structural funds and the new guidelines for the European Social Fund provide an opportunity for the European Parliament and Commission to give a lead in helping member states to develop more cost-effective strategies for tackling long-term unemployment and poverty.

## Key elements in a more cost-effective strategy to tackle long-term unemployment

The work of the projects identifies the following key elements as necessary:

strategies of support for the unemployed, and other groups in poverty, which guarantee a decent level of provision for material needs (income, housing, healthcare, transport, food and fuel, etc.) and co-ordinate necessary support services; and which enhance their autonomy, and counteract those processes which create dependency and reinforce powerlessness;

strategies of investment and economic development which compensate those urban and rural areas from which private and public capital has been withdrawn, leaving behind disproportionate concentrations of the unemployed, the elderly and the poor; and which regenerate the economies of those areas by investment in indigenous skills, technologies, and employment opportunities.

## The need for a guaranteed minimum income

The European Commission has set as its target a reduction of long-term unemployment from its 1988 level of 50 per cent of total unemployment to a level of 30 per cent by 1990. Even if this reduction is obtained, the numbers of long-term unemployed in the Community would remain high, at nearly 5 million in 1990. This high level of long-term unemployment raises a number of policy questions as to what kind of income and social support can be provided to the unemployed and those on the margins of the labour market, who face the prospect of several years living on a low income and long-term social isolation. Many projects are experimenting with measures to alleviate these problems and are providing immediate social support and direct assistance to those out of work. However, measures for social support, skill training, and work experience need to be backed up by a guaranteed minimum income, which recognizes the additional financial burdens and needs incurred the longer unemployment persists (see the chapter by Dyson).

## Poverty traps

A number of projects have identified various 'poverty traps' within the social-security systems which prevent some people from working. These traps affect the homeless, single parents, and the disabled in particular. The problems facing the homeless, most of whom are unemployed, are particularly acute. For example, in the United Kingdom bed-and-breakfast accommodation in which local authorities place homeless families, is paid for by the state only for as long as the adults concerned are not employed. Regulations of this kind are self-defeating in that they perpetuate dependency on state benefits.

## The need to harness grassroots resources

Many of the projects have found that an important source of support for the long-term unemployed is the provision of community-based resource centres. These resource centres provide:

> access to information and advice about financial and legal problems and welfare rights;
> social support and activity to overcome isolation and to maintain confidence;
> links to trade unions, to maintain work-based relationships and solidarity;
> a base for representing the needs of the unemployed, and for generating grassroots plans for economic development.

## The need for high-quality skill-training programmes

Many projects would argue from their experience that the unemployed do not need further incentives to find work; what they need is permanent jobs. They point to the weaknesses in current training schemes which often fail to provide a route to permanent employment. By implication they are, therefore, critical of European Social Fund support for these kind of temporary schemes. They are also critical of the quality of many of the schemes subsidized by the European Social Fund. These criticisms include the quality of the work performed, the participants' opportunities for further qualification, and the potential for participants to stabilize their further employment careers. The experience of the projects could be particularly informative in identifying these kinds of weaknesses and pointing to alternatives.

## The role of the European Social Fund

As far as the European Social Fund is concerned, the experience of the projects tackling unemployment would suggest that less of the Social Fund expenditures should be allocated to temporary employment and training measures, and more to specific high-quality skill-training workshops and projects. These projects and workshops would require:

the recruitment of very high-quality staff;
the provision of very high-quality equipment for training purposes;
creche and child-care support and flexible timetabling to make it possible for parents with child-care responsibilities to undergo training or retraining;
a training programme geared to the development of both confidence and skills;
work-placements in local firms and close links with local employers, to provide an unbroken bridge from training into employment;
help with business planning for those who decide to set up their own small firm or co-operative;
follow-up support to those who have entered employment.

Specific high-quality training of this kind, geared to

local employment opportunities, is exactly the next stage of support needed by the unemployed who have been involved in many of the Anti-poverty Programme unemployment projects. However, it will only act as a bridge to employment if there are appropriate job opportunities within the local labour market. For this reason, the attack on poverty and unemployment requires an integrated programme which combines programmes of social development and skill training with economic development and job creation.

## The role of the European Regional Development Fund and Common Agricultural Policy

The work of the projects concerned with long-term unemployment and with rural underdevelopment has important implications for both the European Regional Development Fund and the Common Agricultural Policy. As argued earlier, the root cause of much of the poverty confronted by the projects in both urban and rural areas is the restructuring of key industries, the run-down or withdrawal of private and public investment, and collapse into a vicious circle of economic and social decline. The restructuring of at least some of the industries concerned (e.g. agriculture, steel, coal, shipbuilding, textiles) has taken place as part of a European process, or even plan. Some of these European restructuring plans have included compensation to firms agreeing to go out of business in order to reduce 'over capacity'. In many cases individual workers from those industries have been compensated through redundancy payments. However, in very few cases has the restructuring been accompanied by a redundancy payment to the community affected by the job loss. Coal- and steel-closure areas have received some compensatory funding, but this funding has not succeeded in generating jobs in sufficient numbers, or of the right type, with sufficient speed to match the needs of those made redundant. Much more integrated plans for redundancy counselling, tailor-made retraining, and local job creation are necessary as part of the process of industrial restructuring, if a long period of unemployment, poverty and economic and social decline is to be avoided in such areas. (5)

## The need for targeted investment and economic development

It is clear, however, that the project areas have gone beyond the stage where preventative action can be taken. Hence strategies have to focus on compensation and cure. Areas of concentrated unemployment and poverty need:

> substantial investment in major capital projects to restore the physical and social fabric of the area and to regenerate the local economy. These projects need to be in the manufacturing, construction, and service sectors and to be built wherever possible on the foundation of indigenous industries, technologies, and skills;
>
> programmes of high-quality skill training to match both the particular needs of the specific redundant and unemployed people in that local labour market, and also the needs of the emerging new businesses;
>
> new organizational machinery, along the lines of a regional enterprise board, to develop the necessary business plans and training plans, to attract and manage the investments, and to provide the product development, marketing, and business development advice.

## The need for a third anti-poverty programme

The above paragraphs have identified some of the key elements in a strategy to tackle long-term unemployment. Many of the elements require a close integration of the contributions from the European Social Fund, the European Regional Development Fund, and the Common Agricultural Policy, with national, regional, and local-government resources. The need will remain, however, for a specific European programme to combat poverty. The 2nd European Programme has demonstrated that locally based action-research projects have been able to help not only in diagnosing the problems, but in developing the solutions. Voluntary and grassroots organizations have played a crucial role in this process and should continue to be active partners in any 3rd European Programme to Combat Poverty.

## 1992 and all that

The late 1980s are the crucial moment for a concerted and integrated economic and social strategy to combat long-term unemployment. The moves towards a single European market in 1992 will undoubtedly lead to a further wave of restructuring in certain industries and technologies. This process runs the risk of generating further concentrations of unemployment and poverty which will undermine the aims of the single market in at least two ways. First, mass unemployment and poverty will cause a heavy fiscal drag on the European Community. The costs of social assistance, lost taxes, etc., will escalate even further, and increase public expenditure dramatically (particularly at a time when the European population is also ageing and pushing up social-security and pension costs). Second, mass unemployment and poverty in effect shrink the potential size and scope of the European market, as the unemployed and poor lack the incomes necessary to consume goods and services.

There are, therefore, strong economic and social reasons for the European Commission and Parliament to give a lead in developing mainstream strategies for tackling unemployment and poverty as part of the moves towards a single European market.

## REFERENCES

I am happy to acknowledge my indebtedness in the preparation of this chapter to colleagues in the European Programme to Combat Poverty. In particular, I have drawn a lot of material from visits and discussions with the projects concerning long-term unemployment and youth unemployment, and with colleagues in the evaluation team. Many of the ideas have been developed in discussion with Jef Breda, Frank Laczko, and Mike Morrisey, and I am grateful particularly to them. I write, however, in a personal capacity and my views do not necessarily represent those of the Commission or of the Animation and Dissemination Service.

1    M.O. Higgins and S. Jenkins (1988) Poverty in Europe, Bath: Centre for Analysis of Social Policy, Bath University, February.
2    G. Room (1987) 'New Poverty' in the European

Community, Bath: Centre for Analysis of Social Policy, Bath University, May.

3    M.J. Piore and C. Sabel (1984) The Second Industrial Divide: Possibilities for Prosperity, New York: Basic Books.

4    J.K. Galbraith (1979) The Nature of Mass Poverty, Cambridge, Mass.: Harvard University Press.

5    See, for example, the new initiative in West Germany (reported in Transition, September 1987) to use training to minimize the number and impact of redundancies in the Hamburg shipyards, before they take place, while linking training to economic policies.

## Chapter Six

## LOCAL LABOUR-MARKET DEVELOPMENT AND LONG-TERM UNEMPLOYMENT IN THE EC COUNTRIES

Haris Martinos and Andrea Caspari

Between 30-50 per cent of the European Community's 16 million registered unemployed have been unemployed for a year or more, and this phenomenon of long-term unemployment has been increasing in recent years. The scale and persistence of unemployment have led to repeated calls - both nationally and on an EC level - for an active commitment and concerted effort to tackle this problem. The heads of governments of the EC member states have identified the long-term unemployed as requiring particular help, and have stressed the need for specific measures that go beyond general unemployment measures (see the chapter by Johnstone). Constraints at the 'supra-local' level have often restricted the scope of macro-economic measures to deal with long-term unemployment. However, necessity being the mother of invention, new policy directions and innovative approaches have been explored to attempt to cope and to reduce unemployment levels.

The emergence of a local approach, characterized by locally generated and/or delivered actions, and with its emphasis on the mobilization of existing local resources, has become increasingly important in tackling these problems throughout the European Community. In part, the growth of local strategies to tackle unemployment has been a spontaneous development: local authorities and other local agencies in the public, private, and voluntary sectors have been confronted with long-term unemployment and have had to find ways to cope with it in their everyday dealings. However, there has been a growing recognition that a local approach to employment and economic development, one which relies on local action for its realization, is a viable policy direction in general and for tackling long-term unemployment in particular. This process of recognition has been reflected in a range of programmes adopted by the Commission of the European Communities.

At the beginning of 1986, the European Commission launched an Action Programme on Local Labour-Market Development. The Programme examined how twelve areas in the European Community have been tackling unemployment and employment development. The areas chosen were all characterized by high levels of unemployment, although in other respects they differed significantly: less developed rural areas were represented, as were rapidly declining traditional industrial areas, and also relatively prosperous conurbations which nevertheless faced considerable employment problems.

This programme of 1986 was not specifically targeted on long-term unemployment. However, it emerged that local approaches to employment and economic development represented a viable approach both in the rural and industrial areas where long-term unemployment was an integral part of deep-rooted problems. It was evident that a local approach had much to offer in terms of the utilization of resources, the development of employment, the stimulation of innovation, and the creation of new partnerships in a community - all crucial steps in a process of sustainable development. It is from the perspective of this programme that this chapter considers the significance of local labour-market development for long-term unemployment.

## LONG-TERM UNEMPLOYMENT: THE GENERAL CONTEXT

Long-term unemployment has various underlying causes and forms and affects different sectors of the population in different ways. In the industrialized/urban areas long-term unemployment has typically resulted from industrial restructuring on the one hand, for example, the introduction of new technologies and capital-intensive production processes. On the other hand, long-term unemployment has been caused by closures of traditional industries in areas dependent on a single industry and employer, such as coal-mining and shipbuilding. In the less developed areas, mostly on Europe's periphery, long-term unemployment can be equated with the endemic seasonal unemployment or underemployment in agriculture and more recently in tourism in rural areas.

By and large, recent concern about long-term unemployment has focused on the former category - that is the growth of long-term unemployment in the traditional industrial areas of Europe - rather than on the less developed areas. In part this is because the change in the former employment centres of the industrial heartland of Western Europe has been more visible, and more dramatic. The chronic employment problems of Europe's less developed peripheral areas have had less prominence, not least in these countries' policies which have tended to focus on the development of the industrial rather than the rural areas. However, seasonal unemployment and underemployment especially in agriculture can be seen as a problem of long-term unemployment; and the scale of the problem can be gauged by the massive scale of emigration from these areas not only in the 1960s and early 1970s, but for well over a century. As the option of migration from these areas has become more limited since the mid-1970s, unemployment has correspondingly risen.

Bearing in mind then that long-term unemployment is a problem facing many parts of Europe and not only the industrial areas, one must ask who has been primarily affected. The main groups of long-term unemployed have been identified as follows. Young workers with insufficient education levels and a lack of basic literacy and numeracy skills, as well as a lack of training, qualification, and/or job experience, have generally been severely hit by long-term unemployment. Older people who have been made redundant and who lack retraining for new skills or who can for other reasons not re-enter the labour market, are another major group of long-term unemployed. Similarly women, members of ethnic groups, foreign workers, and others who may experience cumulative forms of disadvantage are often numerous among the long-term unemployed.

Of these groups, priority in terms of the development of policy has often been given to the young long-term unemployed, special concern having been voiced about those school-leavers who have never had a job. In Greece and Italy, for example, the young long-term unemployed are the main focus of long-term unemployment measures. Unemployment figures for young people are in general two or three times as high as for adults in most of the EC member states.

While long-term unemployment has hit these groups particularly hard, unemployment has become a more widespread feature and has extended throughout society. At the present rate of increase there are projections for 1990 in which it is estimated that about one in eight Europeans could be out of work (Council of Europe Report on Local and Regional Authorities and the Challenge of Unemployment, 1983, p. 15).

Both the scale and the complexity of the problem - in terms of the range of people affected, and the spread of the problem throughout the Community - clearly call for imaginative and committed approaches at all levels.

## INITIATIVES AND APPROACHES IN LOCAL LABOUR-MARKET DEVELOPMENT

There are a number of important reasons which can be highlighted to explain the emergence of local approaches to unemployment and employment development. The seriousness of employment problems, a scarcity of footloose investment, constraints at the supra-local level in tackling long-term unemployment, and consequent disillusionment with the effectiveness of supra-local policies, have all contributed to the growth of local actions and a recognition that these may have something to offer. In addition, an awareness that, despite grave problems, areas may nevertheless have underutilized local resources has strengthened the credibility of local initiatives and programmes throughout Western Europe. Several programmes of relevance to the particular problems associated with long-term unemployment have been created.

Among these is the OECD's ILE Programme (Programme for Local Employment Initiatives) to promote growth and employment initiatives, especially in depressed areas with high unemployment. The objectives of this programme are to promote a wide exchange of information and experience on the development of local employment and enterprise; to design methods for evaluation of projects and undertake the social and economic evaluation of new local enterprise and employment initiatives; and to provide technical support for participants in the programme. Like other related programmes, the ILE Programme has started from the assumption that, even in depressed areas, there are local resources which could contribute significantly to local labour-market development. The Programme has, for example, encouraged educational institutions and large firms to take on new roles in local development and job creation, and it has supported the extension of the role of employment services.

In a similar vein, the value of the local dimension and its complementarity to EC and member state employment policies has been recognized at the level of the European Community. The European Social Fund has identified local initiatives as a priority category, and the European Regional Development Fund is assisting endogenous development. Local integrated approaches to development have been strengthened by the introduction of the Integrated Mediterranean Programmes and the prospect of a new generation of Integrated Development Operations, supported by the EC. Other programmes set up by the European Commission to support locally inspired and/or delivered actions include a programme to foster exchanges between local development agents; a European Information Exchange Network on Local Employment Initiatives (ELISE); and the Action Programme for Local Labour-Market Development, which will be described in greater detail in the following section.

There are questions raised about local initiatives and programmes supporting local approaches as to whether they are mainstream or marginal to unemployment problems, and whether these local approaches and their goals should be seen as short or long term. It is also frequently emphasized that these local actions and development strategies (and the EC/member state policies that cultivate them) should not become stop-gap measures to compensate for deficiencies in national policies; and, furthermore, that they should not filter public funds away from the priority of pursuing permanent jobs and a return to economic growth. Given these qualifying factors, there is general agreement that local initiatives in the field of labour-market development are not attempts to replace mainstream-policy efforts to foster economic growth, but rather measures that are complementary to macro-economic EC and member state policies, as part of a broadening of the fight against long-term unemployment. A range of local initiatives and strategies in the field of labour-market development was explored as part of the EC Action Programme on Local Labour-Market Development.

## THE EC ACTION PROGRAMME ON LOCAL LABOUR-MARKET DEVELOPMENT

At the beginning of 1986, the Commission of the European Communities (Directorate-General V, responsible for Employment, Social Affairs and Education) launched an Action Programme on Local Labour-Market Development, now known as LEDA - Local Employment Development Action Programme. The programme examined how twelve areas in the European Community were affected by unemployment, and how they have been tackling it and employment development. These pilot projects covered diverse areas which had in common relatively high levels of unemployment.

### The areas

The pilot projects were undertaken throughout the EC in declining rural regions or sub-regions, industrial or mixed semi-industrial areas, and larger urban centres, all characterized by unemployment higher than the national average. These were as follows (see figure 6.1):

**Rural**
Les Baronnies, Drome, France
Mid-West Region, Ireland
North Alentejo, Portugal
Province of Sitia, Crete, Greece

**Industrial or semi-industrial**
Aalborg, Denmark
Ravenna, Italy
Tilburg, Netherlands
Le Bruaysis, Nord Pas de Calais, France
Genk, Limburg, Belgium

**Larger urban areas**
Barcelona, Spain
Hamburg, Germany
Nottingham, UK

119

# Figure 6.1 Pilot projects of the EEC action programme

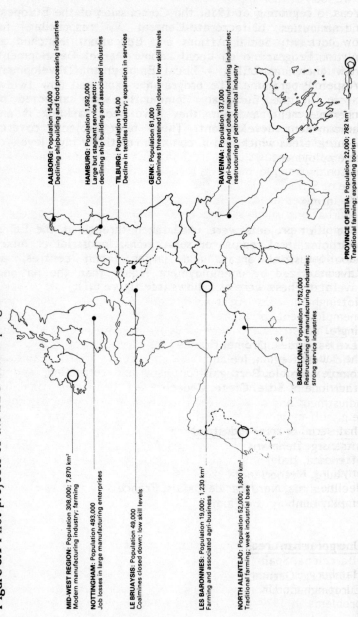

**MID-WEST REGION:** Population 308,000; 7,870 km²
Modern manufacturing industry; farming

**NOTTINGHAM:** Population 493,000
Job losses in large manufacturing enterprises

**LE BRUAYSIS:** Population 49,000
Coalmines closed down; low skill levels

**LES BARONNIES:** Population 19,000; 1,230 km²
Farming and associated agri-business

**NORTH ALENTEJO:** Population 52,000; 1,800 km²
Traditional farming; weak industrial base

**AALBORG:** Population 154,000
Declining shipbuilding and food processing industries

**HAMBURG:** Population 1,592,000
Large but stagnant service sector;
declining ship building and associated industries

**TILBURG:** Population 154,00
Decline in textiles industry; expansion in services

**GENK:** Population 61,000
Coalmines threatened with closure; low skill levels

**RAVENNA:** Population 137,000
Agri-business and other manufacturing industries;
restructuring of petrochemical industry

**BARCELONA:** Population 1,752,000
Restructuring of manufacturing industries;
strong service industries

**PROVINCE OF SITIA:** Population 22,000; 782 km²
Traditional farming; expanding tourism

## The main causes of long-term unemployment in these areas

If we look at the employment situation in these pilot project areas, a number of points stand out concerning the nature and seriousness of local employment problems. They show how persistent unemployment or underemployment is, not only in the less developed rural areas and rapidly declining traditional industrial districts, but also in relatively prosperous parts of the European Community.

In the rural pilot-project areas, the main employment and long-term unemployment problems were attributable to the prevalence of traditional agriculture, as in the North Alentejo, fragmentation of farm holdings, the weakness or lack of an industrial base, and high underemployment or seasonal unemployment in farming and tourism, as in the case of Sitia.

In mixed areas or large otherwise relatively prosperous conurbations such as Barcelona and Ravenna, long-term unemployment problems tend to be derived from the restructuring of specific sectors, such as manufacturing in Barcelona and newer industries such as petro-chemicals in Ravenna.

In the traditional industrial areas, such as Genk, Nottingham, Le Bruaysis, and Hamburg, long-term unemployment problems are associated with the loss of single industries and large employers in the fields of coal, textiles, and shipbuilding; a lack of alternative skills among the labour force, as in Genk and Le Bruaysis; resulting complex problems of restructuring and retraining; and a traditional dependence on these industries making adjustment and a switch to other forms of economic activity (e.g. self-employment) difficult. It is these economic areas that are most strongly associated with the high and intransigent levels of long-term unemployment and its attendant problems, i.e. bleak prospects for a return to anything like previous levels of employment, and spirals of decline making it difficult to attract or generate employment.

## Unemployment problems in selected areas

Hamburg is regarded as one of the prosperous areas in the European Community. It was the region with the fewest problems in 1977-81, in terms of unemployment and GDP.

However, the picture observed more recently through the LEDA pilot project is very different. Unemployment is, at 13 per cent, much higher than the national average; since 1970 the city has lost 33 per cent of its manufacturing jobs, especially in shipbuilding and the food industry; job creation prospects are poor; and in recent years only one new job has been created for every 22 lost.

The experience of another major port city, Barcelona, confirms how widespread the employment problems are throughout the European Community. A city with one of the highest GDP levels per capita in Spain, and important services and attractions commensurate with its status as an Olympic city, nevertheless has an unemployment rate of 19 per cent. In peripheral areas of the city this rate is nearer to 30 per cent.

The Belgian town of Genk in the province of Limburg is an example of deep-rooted employment problems. Genk is a highly industrialized area with a population of 61,500 of which 31.4 per cent are immigrants. The area is now threatened with the closure of its coal mines which employ 6,400 people. This is on top of existing serious problems in the local labour market: high unemployment (22.5 per cent) with an exceptionally high proportion of long-term unemployed (60 per cent); and low skill levels. 72.8 per cent of the Belgian-born unemployed and 94 per cent of those of immigrant origin have received, at most, lower secondary education.

In less industrial areas, employment problems can be equally fundamental and of a long-term nature, as in the case of the province of Sitia in Crete (population 22,000). In this remote part of Greece, 70 per cent of the economically active population is engaged in agriculture. There is serious underemployment in this sector: the main crops and farming methods are traditional, holdings are very small and fragmented, the terrain is mountainous, and the productivity of farm units is very low. People have been abandoning farming for many years, but the large-scale emigration of the 1960s and 1970s has stopped. This is now putting increasing pressure on the smaller secondary and tertiary sectors, where unemployment is already high (estimated at over 12 per cent) especially among young people.

North Alentejo in Portugal is facing similar problems. This mainly rural area with a population of some 52,000 is characterized by a lack of infrastructure, a lack of skilled professionals, emigration, ageing and illiterate farmers, and youth unemployment.

## Local responses to unemployment and to local development

The pilot projects had the opportunity to look into a wide range of local actions that had been developed to cope with these problems, each reflecting the needs and priorities of the particular locality. It is possible to identify some key strategic directions and similarities in substance of local actions despite the diversity of local initiatives throughout the European Community.

Elements of this common core of initiatives in local labour-market development are as follows:

youth training and work experience programmes;
measures for the long-term unemployed;
support for the development of small and medium enterprises;
measures encouraging innovation and technology transfer.

It is helpful to highlight some of the variations on these key features of local approaches below.

The port city of Aalborg in North Jutland, Denmark - 154,000 population - has been suffering from the effects of restructuring in its shipbuilding and food-processing industries. Its shipyard recently made 1,300 people redundant, and unemployment is consistently higher than the national level. Local actions have been trying to stimulate new industrial growth but the main thrust has been in local labour-market management. Local agencies such as the local manpower board, the local employment service, and the municipality co-operated closely on a number of temporary employment and training programmes. The emphasis here was on creating extra places, involving activities not already performed, particularly for the young and the long-term unemployed. In an area with 8,700 unemployed, the scale of these measures was substantial: thus 2,397 long-term unemployed were placed in work in 1985 under a special scheme; 2,145 apprenticeships were subsidized in 1985; and there were 463 places in 1986 in the municipality's training, work experience, and employment projects, including construction, metal work, and gardening services for old-age pensioners.

In Le Bruaysis (France), the emphasis is similarly on training, although from a different perspective. The main thrust is on rebuilding the human resources which have been devastated by the closure of the coal mines which previously dominated every aspect of local life. Closure not only created a shortage of jobs but also left the area demoralized. The population is ageing, while young people see no future in the area. Some 50 per cent of the 15-20 year olds have never had a job; 68 per cent of the labour force is unskilled; and local firms are facing a shortage of skilled labour. In Le Bruaysis an intermunicipal training association was established, the broad aims of which were to create new skills, change mentalities, and generally upgrade the local labour force. Local projects forming part of this include a local research and studies centre on skills and employment, a centre for the assessment of individual training needs, and a unit on training needs in the construction industry.

In Genk, Belgium, the area's response to the closure of mines has been in the field of developing small and medium enterprises. The local authority, large firms, banks, and university jointly launched actions which aimed to overcome the area's lack of an entrepreneurial tradition. A business centre was opened; a business-advice and training programme was offered; and a project sponsored by major local employers and other organizations provided assistance to small and medium enterprises to improve quality of products and management skills, so that these small and medium enterprises could become suppliers of large local firms.

In Nottingham, UK, agencies have similarly pursued local employment objectives through joint ventures. There has, for example, been a sectoral initiative in the city's vulnerable textiles industry; various bodies including private firms, trade unions, and the polytechnic have participated with the local authority in a range of projects in the textiles sector (e.g. providing consultancy services to dyeing and finishing firms, establishing starter units in the fashion/textile sector, providing training for workers, and setting up a fashion centre).

The Mid-West Region of Ireland, traditionally a largely rural area, has a long-standing industrial strategy, comprised of several strands such as the development of small and medium enterprises, as well as the introduction of new technologies and high-skill training, to supplement the attraction and development of high-technology industry. A new policy dimension is the promotion of advanced business and financial services. In addition, a range of local development activities have been undertaken through a community enterprise programme.

Similar in its emphasis on technology transfer and innovation is the case of Hamburg, Germany. Hamburg has followed a collaborative strategy for technology transfer and innovation with the active involvement of the city administration, the Technical University, the private sector, community enterprises, and chambers of commerce. The technical university provides the focus for many activities, for example, an Institute for Technology Development (founded by private initiative in association with the university), a Consulting Centre (a public foundation), and a Working Group for Technology Development, bringing together some 20 organizations involved in consultancy.

Les Baronnies in France represents an interesting example of a local approach to development in a mountainous rural area. Here an intermunicipal local development association was established with the dual objectives of maintaining and modernizing agricultural activities, as well as solving the crucial problem of population retention. The main thrust of this association's activities was a Young Farmers Programme which provided training, technical and financial assistance, information and support services to some 200 young farmers between 1978 and 1985. This programme involved a package of 14 activities and was operated by a team of animateurs and agricultural advisers; it showed considerable success in achieving its aim of stemming the exodus, especially of young farmers, from the area.

In some of these cases, the long-term unemployed were the focus of specific measures, for example, temporary work-experience schemes in Denmark, or training programmes in France. In other areas the local approaches to employment problems did not tackle long-term unemployment as an isolated issue or treat the long-term unemployed as a separate category, but as an integral part of the locality's difficulties. Particularly in the rural areas, an integrated approach to the area's development (as in Les Baronnies, France) was considered the best way of coping with the area's cumulative disadvantages, long-term unemployment among them.

125

## A LOCAL APPROACH TO EMPLOYMENT DEVELOPMENT

The LEDA Programme sought not only to pinpoint these key characteristics of local approaches to labour-market development in its pilot-project areas, but to stimulate and support local actions in a number of ways. In undertaking the pilot-project studies an interactive approach was adopted vis-à-vis local and regional organizations. Through discussion groups, local consultation conferences, interviews, and other research an effort was made not only to identify the key issues, and in some cases to act as a catalyst between the driving forces of a locality, but also to sharpen up the tools for more effective action. A series of review papers on subjects such as the evaluation of local programmes, and on gaining access to/utilizing resources, was undertaken to contribute to the development of good practice and methods for the formulation, implementation, and evaluation of local employment-development projects, for use both by local actors and by policy-makers.

The LEOA Programme sought both to examine the operation of local labour markets and to elaborate - and disseminate - techniques for more effective approaches. The programme explored strategies in the pilot-project areas for forward-looking labour-market development, local employment initiatives, and various forms of local economic development. It considered the effectiveness of measures, their priorities, and how actions were organized. Through the twelve pilot projects the programme found that local actions are widespread, well-established, and substantial in scale.

These actions are characterized by a 'local' content and may be locally initiated and inspired actions (of which there are growing numbers), or they may also be supra-local measures with a local dimension in terms of their design or implementation. Some of the principal strengths and attractions of local approaches to labour-market development derive from the closer knowledge that exists of local potential as well as local needs. This is associated both with greater feedback possibilities, and therefore greater flexibility and effectiveness, and with greater targeting possibilities. In these respects local labour-market development can also be a useful, cost-effective and relatively rapid delivery mechanism for national policies.

An additional essential ingredient of growing local responsibility for labour-market development is the increasing participation of or co-operation between existing local actors in the field of employment creation. Local and regional authorities, large enterprises, trade unions, voluntary groups, universities, chambers of commerce, and the unemployed themselves have often found new roles in the development of the local labour market. There have also been major changes in the local operation of national agencies, for example, in an expanded local network and/or greater autonomy of the offices of national employment services. New agencies have also emerged throughout the EC as part of local labour-market development processes, namely local enterprise boards or trusts, special units within the local government authorities, intermunicipal bodies, development agencies, community businesses, co-operatives, innovation centres, and private-sector initiatives. Needless to say, this proliferation of activities and organizations in the field is not without its own problems, mainly the duplication of efforts and competition for scarce resources.

Despite the diversity of local approaches, there are additional striking common elements in much of this activity, elements which might be said to constitute the cornerstones of a local approach to employment and economic development. The first feature is the endogenous nature of much of this developmental effort which focuses on the utilization of local resources - human, economic, environmental, cultural, historical, geographical, and so on. The second feature is the entrepreneurial approach of the principal local actors, which means an initiative-taking attitude and a rejection of a dependency mentality, as well as an efficient and inventive use of local resources. Many solutions to employment problems are formulated and promoted from within the localities, although Genk and Les Baronnies, for example, both demonstrate a lack of entrepreneurial strengths in the limited business sense. The emphasis on operating flexibly, through networks of local agencies, represents the third feature - the high degree of local responsiveness and institutional flexibility. Extensive partnerships between the local social partners and a number of government levels in Genk confirm this point. A clear orientation towards an employment-development goal but frequent reliance on intermediate objectives (e.g. the provision of training or the introduction of business 'incubator' centres) illustrates a fourth feature, namely a strong action bias in local actions.

These common features transcend a multitude of

ideological viewpoints and political priorities associated with individual programmes and initiatives on employment development. They suggest a coherent approach to employment development with a clearly local orientation, especially in terms of processes and institutional structures.

Clearly, however, certain dynamic elements which are crucial in a sustainable development process - such as a capacity for innovation in technology and other fields, availability of skills, accessibility to markets, and a potential for local investment - are not always present in a locality. In most areas some of these elements are deficient or exist only in a latent form, or local resources are insufficient. Moreover, strong external forces can be at work. Economic and employment development in such cases is unlikely to be totally self-stimulated and self-resourced. External inputs will be needed, but these are not necessarily incompatible with a local approach. A local approach can contribute in these circumstances by creating some of the positive prerequisites and attributes needed for development, e.g. through training or enterprise development; this may create the conditions for a locality to benefit from outside inputs. A vital aspect is that a local approach can mobilize the wider community over a longer period of time, and there is likely to be a greater commitment from many sides to a local orientation, which has important implications for the long-term goal of sustainable development - and therefore for long-term unemployment.

## ORIENTATIONS AND OBJECTIVES OF LOCAL APPROACHES IN TERMS OF LONG-TERM UNEMPLOYMENT

As we have indicated, in terms of both methods and goals, the local labour-market development approach in various respects lends itself to an application to the problems of long-term unemployment. Because of a greater targeting capacity in local approaches to labour-market development, a greater responsiveness to and knowledge of local needs, and other elements referred to above, it would seem that the range of local actions are eminently suited to tackling long-term unemployment. Moreover, the kinds of actions undertaken as part of local labour-market development strategies - for example, in the fields of employment development, the support of small and medium enterprises, an emphasis on training, the encouragement of innovation through a variety of measures, and more generally the effort to regenerate depressed areas - very largely correspond to national and EC policy objectives and measures for the long-term unemployed, in both the economic and social spheres.

National measures and policies (both short and long term) targeted specifically for the long-term unemployed tend to fall into several categories: measures to create new work, measures to improve the 'employability' of the long-term unemployed, and measures to improve the general quality of life for the long-term unemployed. In all of these spheres there is an affinity between national policy objectives and many of the actions already outlined which form the core of local strategies.

'Enterprise-oriented' measures to tackle unemployment, or measures to create new work, are found in many of the EC member states. These measures are often based on an underlying assumption of self-help, for example, through support for small and medium-sized enterprises; provision of counselling and advice services for those wishing to set up in small businesses; and provision of information and resource centres. These elements also often form part of local employment strategies.

'Work-oriented' actions for the long-term unemployed are also common throughout the EC in member states' policies. These latter groups of actions tend to be geared towards a return to work. The assumption in many cases is that the long-term unemployed need help to improve their job prospects, and to maintain or increase their ability to compete for jobs - or at least reduce their alienation from the world of full-time work, for a time when employment availability improves. Actions here might be new or temporary job-creation measures; work-experience programmes, or other forms of reintegrating the long-term unemployed in work; encouraging the long-term unemployed to participate in voluntary work projects and schemes; and actions aimed at employers, such as financial incentives to recruit the long-term unemployed.

Training is another facet of work-oriented activities for the long-term unemployed, i.e. in which there is also (often) an assumption of an ultimate return to work. This training may take the following forms: vocational and skills training for women, or for the long-term unemployed who have never had a job; retraining for new skills of long-term unemployed workers made redundant from traditional industrial sectors in decline; provision of basic educational qualifications, such as literacy and numeracy skills; and apprenticeships.

As we have seen, these existing elements and priorities of national policies for the long-term unemployed are also core elements of local labour-market development strategies in some of the pilot-project areas, although they may not focus exclusively on the long-term unemployed. There is a similar affinity between local and national priorities in employment development also in the social sphere. The above-cited measures focus broadly on stimulating the employment of the long-term unemployed and on economic regeneration. Another important area of activity for the long-term unemployed is that which has more of a 'social' or 'community development' focus. These are less explicitly 'economic' actions than ones whose objective is to improve the quality of life of the long-term unemployed. Actions in this sphere may range from provision of health advice, personal counselling, child care, and leisure facilities to cultural projects of various kinds, and a long-term emphasis on a regeneration of depressed communities that goes well beyond the 'economic', both on the individual and the community level. This concern for providing for the social welfare of the long-term unemployed and for community development is likewise thoroughly compatible with other local economic-development objectives and actions.

## CONCLUSION

To summarize, there exists within the body of local labour-market development actions and strategies in the EC a series of actions which, whether targeted or not on the long-term unemployed, is highly relevant to the problems associated with long-term unemployment, and highly compatible with macro-level objectives and priorities in this sphere. Three main strands of actions for the long-term unemployed, or models of best practice, can be singled out from the diversity of local labour-market development activities.

> Overall broad and relatively untargeted development strategies and processes which simultaneously seek to

'upgrade the human potential' and to create employment, benefiting the entire area including the long-term unemployed.

Targeted actions for the long-term unemployed which form part of integrated programmes (e.g. Les Baronnies - Young Farmers Programme).

Targeted short-term measures such as 'recycling' the long-term unemployed through training, temporary work programmes, etc.

In these cases it must be borne in mind that, though the targeting capacity of a local labour-market development strategy may allow a focus on the long-term unemployed, this focus cannot be sustained in isolation and cannot be an alternative to the objective of generating overall economic development (see the chapter by Whyatt). And the indications are that, in order to achieve this objective of economic and employment development, a locality needs to operate within a supportive context. It appears that a number of steps taken by the European Community and the member states would contribute to the enhancement and greater utilization of the strengths of the local approach as a complement to other employment policies. Member states could, for example, be encouraged to guarantee a minimum level of devolution to local agencies for the design and implementation of actions supported through national schemes; and they could establish a clear legal and financial framework concerning regional and local government actions on employment development, reflecting the greater level of local autonomy in the operation of national agencies suggested above. It is within such a context that member states' employment policies could be more effectively implemented, and that local agencies and actors would be more able to realize their objectives of tackling unemployment and cultivating a sustainable development process.

**Chapter Seven**

## THE ROLE OF LOCAL EMPLOYMENT INITIATIVES IN COMBATING LONG-TERM UNEMPLOYMENT IN THE EUROPEAN COMMUNITY

Tomas Roseingrave

The role of locally based initiatives or local employment initiatives (LEIs) as agents contributing to the amelioration of long-term unemployment must be seen and evaluated within the constraining framework of the wider global economic and financial system. No nation state, or even the European Community, much less a local community, is any longer a self-contained economic island. Moreover, the opportunities for effective locally based initiatives, involving appropriate participation by people in decision-making and access to the utilization of resources, depend on the measure of freedom permitted to them by the national government (see the chapter by Whyatt). Some of the member states of the European Community (including Ireland and the UK) suffer from increasingly centralized political systems. Nowhere is this process of centralization more evident than in the administration, use, and distribution of the financial instruments of the European Community, notably the Regional Fund and the Social Fund, in which local authorities and local voluntary development agencies play little direct part. Finally, few countries within the European Community have provided adequate legislative support for LEIs by way of specific laws or favourable legal regulations.

Against the background of the limitations imposed by these three major factors at global and nation-state levels this chapter analyses from two perspectives the role of locally based initiatives in assisting the long-term unemployed in the European Community:

1    Their role as evident from surveys that have been made

132

over the last four or five years and also from personal experience of working over a number of years with local voluntary development groups. The potential of LEIs is recognized as a worthwhile contribution to combating the rising tide of unemployment, especially long-term unemployment, on the European labour market where, by 1987, of the 16 million people registered as unemployed, some 7 to 8 million had been unemployed for a year or more. (1) Some indicators suggest that one-third of these 7 to 8 million have been unemployed for more than two years. Although there are, of course, different geographical concentrations of this unemployment, these figures represent the overall magnitude of the problem of long-term unemployment.

For an operational definition of local employment initiatives (a term also used by the OECD), the European Commission has provided the following:

> those initiatives that have occurred at the local level - often involving co-operation between individuals, action groups, the social partners, and local and regional authorities - with the specific aim of providing additional permanent employment opportunities through the creation of new, small-scale enterprises. (2)

(One might have reservations about the validity of using the term permanent employment in this context, especially in the light of the effects of rapid technological innovation on employment.)

So what then are the potential, the problems, and the limitations of LEIs in helping to combat long-term unemployment in the European Community labour market, viewed from their operation at the local level? That is the first perspective in this chapter.

The role of LEIs has been discussed, debated, and analysed by the representatives of the social partners in the Economic and Social Committee of the European Community. The Economic and Social Committee is a consultative assembly of 189 members drawn from the major European employers, trade unions, farmers, and professional organizations and derives its mandate from the Treaty of Rome. It set out its views on LEIs in two Opinions: one by way of response to proposals for action by the European Commission relevant to the design of strategies and programmes to combat unemployment (see the chapter by Johnston); and the second was an initiative of the ESC itself. (3) This second perspective on the contribution of LEIs in helping to combat long-term unemployment is, therefore, from the European level in particular, the viewpoint of the social partners as it has emerged from the discussions and analysis within the ESC.

## THE NATURE, ROLE AND DEVELOPMENT OF LOCAL EMPLOYMENT INITIATIVES IN THE EUROPEAN COMMUNITY

Local employment initiatives in the European Community operate under diverse structures and legal models - workers' co-operatives, companies owned by the people of a local community through some representative structure such as a local community council, companies in private ownership, companies owned by a sectoral-interest group within the local community, non-profit making companies, and so on.

In the 1980s there has been a rapid growth of LEIs. They are usually small in size, comprising from three to four people up to 15 to 20, and are established in both urban and rural areas. However, they can employ 100, 150 or 200 people as is the case in several workers' co-operatives in Ireland, manufacturing pottery, footwear, furniture, etc. As private investment tends to decline in localities and regions affected by economic decline and a lack of job opportunities, the LEIs seek to substitute limited endogenous development through locally generated employment which is related to regionally and locally identified needs and resources. Since they place the provision of employment as their primary objective and seek to achieve economic viability in difficult market conditions, LEIs are often obliged to accept relatively low pay rates for unduly long hours of work. Understandably, this situation left trade unions less than enthusiastic about them certainly in the early years of their development.

At the same time it must be remembered that LEIs represent a different approach, one that does not fit too easily into the accepted schema of economic organization and development. This approach to the creation of employment opportunities often requires the temporary foregoing of personal short-term benefits in pay and conditions of work in order to ensure that neighbours and colleagues can share in some of the benefits accruing from

134

employment and from the human dignity of work. Behind LEIs lies the implicit assumption that we are all, personally and communally, responsible for the welfare of the fellow citizens of our community and its future development. Such an approach must, however, always have regard to the hard-nosed realities of the market-place.

It is estimated that more than half a million people are employed in workers' co-operatives throughout the European Community. According to the European Commission, on the basis of its research and consultations, it would seem that 'a considerably larger number of people are involved in other enterprises which could be categorized as LEIs'. (4) The employment of half a million people by LEIs seems a relatively small contribution to combating the huge figure of 16 million unemployed in the European Community (or the lower figure of 8 million long-term unemployed). Measured in terms of direct employment creation, the contribution of LEIs to the amelioration of Europe's massive unemployment problem is relatively small and modest. Their role must be seen as complementary to, and not as a substitute for, macro-economic and structural policies and programmes at the national and EC levels. LEIs are a modest but limited direct contribution to a major European problem that is influenced by global trends and events.

However, assessment of the contribution of LEIs cannot be confined to their limited capacity for direct employment creation. Their really effective role occurs by definition in the localities and communities where they have been established. After research and consultations in the member states of the EC (mainly through the Centre for Employment Initiatives (CEI), London, under the direction of Peter Kuenstler), and in discussions with the workers' co-operative group at the European level (CECOP), and in close collaboration with the OECD's Co-operative Action Programme on Local Employment Initiatives, (5) the European Commission emphasized the developmental role and potential for cumulative benefits that LEIs possess, an influence that extends well beyond their own walls. In certain deprived or declining areas LEIs may be the only source of new employment. According to the European Commission, drawing on the experience of their operation within the member states of the EC, LEIs have this potential and influence because

they embody and promote ideas such as self-help, co-operation, local and regional regeneration, or support for the disadvantaged which can have a positive influence on the thinking and behaviour of other enterprises and public bodies. The cumulative benefits can be considerable, particularly in areas in which the endogenous capacity for economic development has been lost, often due to over-dependence on one or two large industries or firms that have gone into decline. LEIs can help revive this capacity and restore confidence by providing legitimate vehicles for using local energies to meet local needs, avoiding them being diverted into 'black economy' activities or dissipating in other ways. (6)

On the other hand, for their effective launch and development, LEIs need a supportive local community milieu. This support depends on the efforts of an individual, group, or agency to change social attitudes towards openness to new ideas and to instil the motivation to take effective steps towards change and development. Such a transformation is not always easy to achieve, particularly if local culture and social mores are ignored. A study of successful local community-action programmes aimed at employment creation in Ireland illustrated that a key criterion of success was that LEIs must be seen to be rooted in the ethos and way of life of the people of the local communities. (7) They depended on a culture in which people were bonded by common ideals and a memory of shared experience. There was a local culture and a core of attitudes that influenced the daily lives and perspectives of the people. In developing LEIs great care has to be taken to ensure that what is technically possible, and appears socially advantageous, does not inadvertently conflict with established cultural attitudes and social patterns. Economic strategies for local development must be reconciled with the values and life styles of the people. In other words, innovation has to be integrated with cultural continuity.

Another Irish study of entrepreneurs highlighted the importance of the social, economic, and psychological factors involved in setting up new ventures. Unity of action and co-operative efforts provide strength and conviction necessary for achievement and the development of a favourable local environment. On the other hand, fragmentation of local effort has too often resulted in failure. This factor was specifically identified in the series of consultations on local employment initiatives held in the various member states of the European Community and contained in a report (No 02/84) of the Centre for Employment Initiatives. Of particular interest is the section on the consultation in Fraserburgh, Scotland (3-4 December 1984), in which it is stated that:

> There was a realization that there existed in the local community a problem of fragmentation. It was felt that by coming together they could not only create a louder voice in lobbying outside bodies, but could also establish a general development plan for the community. (p. 32)

It is important to take note of the hierarchical levels in socio-economic development and in the creation of employment opportunities through LEIs. Local community initiatives must, of necessity, have regard to the policies, plans, and programmes of the national government, of the local authorities, of the statutory agencies appointed to implement the programmes, of the regulations governing the operation of the financial instruments available at the EC level (i.e. the Regional and Social Funds), and of the role and function of various industrial development and tourism development agencies charged with implementing socio-economic programmes. On the other hand, if the governmental system is characterized by excessive centralization and undue bureaucracy (referred to at the beginning of this chapter), little can be done (except by faint praise or lip-service on appropriate platforms) to encourage and promote the development of the core values of responsibility - personal and communal - within local communities and to exploit communities' potential to contribute to their own economic and social development and the development and prosperity of the society.

If a devolution of power does not characterize the political and administrative systems, the tasks of local authorities and local communities in trying to play their appropriate roles in the hierarchical levels of development become increasingly more difficult, frustrating and stultifying. The centre of administrative power, in London or Dublin, will only too often decide both what is best by way of development for all those regions and localities outside the pale of the central administration and how much of the available financial resources will be allocated to such development as the centre permits. Too many member states of the EC place excessive powers of decision-making and the allocation of the financial resources in the hands of central departments of state; the latter largely determine the projects and the areas that will benefit, for example, from the ERDF. Unfortunately, the regulations governing the ERDF make it possible for central governments to centralize decision-making about the distribution of regional aid.

This centralization of decision-making about the identification of projects and about the allocation of EC Fund monies breaches also the principle of 'additionality' in the regulations of the ERDF; this principle means that the development projects are intended to be, and should be, in addition to regional development projects already provided for in the programmes and budgets of the member state. The principle of 'additionality' is 'legalistically' and effectively circumvented by the process of claiming back the monies of the ERDF into the coffers of the central exchequers of the member states. In effect, in many cases no additional plans or projects are provided for the regions and their local communities to carry out local initiatives directed at the creation of local employment. Most LEIs have been budgeted for already, and their cost is then recouped from the ERDF.

Over the years the Economic and Social Committee of the European Community has argued for a change of this procedure; the European Parliament did so in November 1987 by way of a unanimous resolution. For both the aim was to enable local communities, through their regional authorities, to be directly involved in the necessary decision-making to identify development projects and to allocate the relevant resources. Several benefits could accrue. It would enable people and their regional development institutions, in the spirit of 'participatory democracy', to be involved in the relevant decisions for socio-economic development that are likely to affect them, their families and their neighbourhoods; and it would encourage a greater sense of identity between the goals and expectations of the people of the local communities and the decision-making processes and functions of the institutions of the European Communities which are carried out in 'remote' cities of Europe - Brussels, Luxemburg or Strasbourg. In conclusion, within the hierarchy of processes of socio-economic development, the lower levels - the local community and the local and regional authorities - must be

enabled to play their full and proper roles in the identification of their needs and to bring their available local
resources into the sum total of the developmental assets.

A recent study, in which ten local initiatives in Ireland
and twelve in the region of the Emilia-Romagna in Italy
were studied and analysed as part of the first FAST
programme of the European Commission (Programme of
Forecasting and Assessment in Science and Technology),
found that the role of financial institutions in venture
initiation also needs to be clearly defined:

> Understanding of the application of the various
> entrepreneurial support schemes and entitlements was
> perceived as a difficulty. Administration of grants and
> loans needs to be streamlined ... the institutional
> process of project evaluation was not fully understood
> and did not encourage assumption of responsibility at an
> early stage by potential business owners/promoters.
> More attention needs to be given to the business owner;
> at present the emphasis is on the venture. (8)

Together with certain legal and fiscal obstacles a lack of
adequate local and national financial resources is an
inhibiting factor in local employment initiatives.

In the report on the second series of local consultations
carried out for the European Commission by the Centre for
Employment Initiatives (London), the problems and
difficulties encountered by local groups that sought to set
up small businesses of any kind (traditional or non-
traditional) were discussed. 'With regard to finance,' says
the report (p. 74), 'banking institutions were seen as
unsympathetic to the needs of LEIs.' While banks could be
much more sympathetic and responsive, they are often
'unsympathetic' to local initiatives because the groups lack a
legal-entity status. This problem can be overcome by the
groups establishing themselves within some legal category
such as a friendly society, a co-operative, or a limited-
liability company.

Likewise the need for a change in the legal systems in
the EEC to accommodate the new-style enterprises was
indicated in a report to the European Commission by
Holland, Daviter, and Gessner (1984). They said:

> Existing commercial, labour and fiscal law is not
> attuned to the aims and organizational types of these

139

small 'post-industrial' firms. The priority of meaningful work over increasing wealth, self-administration without a permanent hierarchy, common utilization of assets without the possibility of individual access or exploitation, an inclusive concept of work which also includes elements regarding the common interest - these are the aims, which are hindered rather than upheld by the law and the legal practice based on it. Adaptation and alterations are above all necessary in legal forms ... (9)

Another important factor inhibiting the development of LEIs at the local level is a failure to communicate information about available schemes. Problems of knowing where to go and whom to ask for advice and information are endemic in the work of local groups. The publication of booklets, pamphlets, brochures, and documentation of all sorts, however excellently produced and presented, is not in itself sufficient to ensure that the advice and information is communicated and has seeped down to the local community levels. Communication is an old problem in social services - health, social welfare, education, employment - and in giving effect to the rights of citizens to such services. Obviously, if meaning is to be given to the concept and practice of democracy, such services must be brought much closer to the appropriate local level where they can be properly delivered to those for whom they are intended. Local information centres, such as the Community Information Centres piloted by Muintir na Tire in Ireland, are needed to overcome the problem of communicating from central agencies to local groups. At the European level we now have ELISE (located in Brussels) - the European Network for Information Exchange on Employment Initiatives. This organization is helping to fill the information gap at the EC level and issues a monthly newsletter which incorporates selected information from the Official Journal of the EC, the EC press agencies, and the European Parliament's press review on a wide variety of topics relevant to LEIs - employment-creation initiatives, economic and social development activities by local authorities, projects launched by district associations, trade unions, or local groups. ELISE aims to facilitate exchanges of information throughout Europe by means of studies, project filecards, an information network, and an ideas forum.

## THE EUROPEAN COMMUNITY PERSPECTIVE ON LOCAL EMPLOYMENT INITIATIVES

The role of LEIs has been discussed, debated, and analysed by the representatives of the social partners at the European level in the Assembly of the Economic and Social Committee, and their assessment and recommendations given in two Opinions in 1984 and 1986. In an accompanying report to these Opinions the Economic and Social Committee outlined examples of the development of LEIs in the European Community. (10) This report was based on material provided by the European Commission and from the members of the study group on data derived from government departments of their own respective member states. The data is meant to be indicative and by no means an exhaustive statement of development of LEIs. A summary of the main findings will be helpful.

## Belgium

Development of LEIs in Belgium has occurred in three main ways in the 1980s: as limited state-sponsored initiatives; as self-managed co-operatives and 'alternative' enterprises; and as worker-producer co-operatives. By 1982 some 51 'alternative' enterprises were founded in Wallonia - 9 in the building sector, 8 in industrial manufacturing, 4 in handicrafts, and 28 in services and transport. The number of jobs created totalled 724. By 1983 'nouvelles co-opératives' alone accounted for 1,500 jobs. Trade unions provided help and advice. Also, in the less depressed Flanders region some 50 enterprises which concentrated on wholefood production and distribution had been recently set up and created about ,000 jobs. The report concluded on the section on Belgium: 'Overall there is clearly a need to tighten up the management of the schemes'. (11)

## Denmark

LEIs have developed in two main areas: as enterprises set up by groups of workers whose jobs were at risk through failure or closure of traditional undertakings; and as part of an 'alternative' cultural/protest movement. The first group was helped considerably by the federal co-operative body (Det

Kooperative Faellesforbund - DKF) which launched a number of schemes involving workers, trade unions, and local authorities. The second category of LEIs - the 'alternative' local employment initiatives - is not so much a response to unemployment as an expression of dissident protest by minority groups which have their origins in the 1960s and represent 'post-materialistic' lifestyles. There are several thousand 'collectives' or 'communes' and alternative entrepreneurial or co-operative projects in Denmark. They vary considerably in size and form, operating as self-governing communities or as small limited companies.

## Federal Republic of Germany

The development of LEIs in West Germany is characterized by the part played by churches, charities, and/or local communities. The Protestant Church, for example, runs about 70 employment initiatives for training up to 3,000 'hard-core' unemployed persons. Local communities have also launched publicly funded initiatives.

The numbers of LEIs created as part of the 'alternative' socio-cultural movement is on a more significant scale, involving an estimated 104,000 persons in some 14,000 'alternative'-sector projects which are self-managed and organized on a collective basis (for details, see the chapter by Kaiser). Work is normally labour-intensive and in small-scale units - 16 per cent are strictly production-based; 25 per cent concern the media, cultural projects such as theatre, dance, and circus; 18 per cent provide alternative social, medical, therapeutic, and educational services; 10 per cent provide leisure facilities; 19 per cent are concerned with 'bio-foodstores'; and 12 per cent provide repair professional, and transport services. An estimated 35 per cent of employees were not previously regularly employed and the average life-span of an 'alternative' LEI was at least five years. The trade unions have also helped to promote local and regional initiatives for job creation.

## France

LEI activity in France is mainly divided into two categories the role of the long-established workers' co-operative body the Confédération Générale des Sociétés Ouvrières d

Production (CSOP); and the newly emerging boutiques de gestion. By 1984 about 1,200 workers' co-operatives of different kinds had been set up in France, employing 36,000 persons. From 1981 under the Socialist government the boutiques de gestion gave a new impetus to LEI development. Their aim was to help create jobs by promoting new forms of local businesses together with local infrastructural development, local social services, and technologically up-dated local workshops. By 1983 some 1,405 companies employing 3,405 persons had been promoted in this way. The majority of new local jobs created were in commerce/services (35 per cent); industry (30 per cent); crafts (21 per cent); construction (10 per cent); and agriculture (4 per cent). The failure rate of such initiatives was estimated at about 16 per cent.

## Greece

Till recently there was not much development of LEIs in Greece. However, the Socialist government has tried to overcome these problems by setting up the Hellenic Agency for Local Development and Local Government and a state holding company, ETEPAP, for LEI projects. The first offers technical support to local authorities and local enterprises in developing economic growth and employment. ETEPAP aims to promote local productive initiatives by young people. It operates as a subsidiary of the Hellenic Industrial Development Bank, and its management is appointed by the Ministry of Youth and the Ministry of National Economy.

## Ireland

Multi-purpose co-operatives began to emerge in the 1960s. Even so, the self-help national Community Development Movement, Muintir na Tire, has been active since 1937 and has helped to organize many and varied LEIs which were engaged in turf/peat-cutting schemes, group water schemes, small industry development, and several others. The primary role of the organization is as an 'animateur' or 'prime mover' group, motivating the people in local communities to exercise personal and communal responsibility through their own activities in representative community councils for socio-economic development. Such activities are always to be integrated with national, regional, and local policies and programmes. Further impetus to LEI development has come from considerable, though different, schemes administered and initiated by the Youth Employment Agency (YEA), Anco (National Manpower Training Authority), and the Manpower Services of the Department of Labour. These three agencies were amalgamated on 1 January 1988 under the new title of FAS.

## Italy

Development of LEIs in Italy has mainly occurred in the form of co-operatives, in keeping with a well-established tradition. Italy has the largest co-operative movement in the European Community. More than 67,000 co-operatives of all types were set up between 1972 and 1984; among these were 14,000 productive worker co-operatives employing 430,000 persons. Little more than half of this number of co-operatives was officially registered. Registered co-operatives represent the majority of truly operational LEIs in Italy. There is a strong liaison between the trade unions and the co-operatives, often resulting in joint projects.

As a result of a law (the 285 law) in 1977 which provided L1,310 billion over four years for youth employment and LEIs, nearly 7,500 co-operatives were established. However, because they were set up very often by people with little experience, these co-operatives have had a high failure rate.

## Luxemburg

In the 1980s governmental measures have been adopted to promote LEI development, especially to help the social integration of young people without any formal education or training qualifications.

## The Netherlands

LEIs in the Netherlands are divided into three main categories: small businesses, local/worker co-operatives, and 'alternative initiatives'. Certain common characteristics can be found in all three categories: the initiatives spring from the local community; they are geared to the local potential; they are often supported by private firms/ organizations; the participants help in the form of money, knowledge, advice, etc.; local trade unions are becoming increasingly involved; and they aim not only to create new jobs but to retain existing jobs. As in West Germany (see Kaiser's chapter), the 'alternative' movement has a significant input into LEI projects. By 1983 an estimated 1,600 'eco-enterprises' (self-managed, small-scale models) were operational. They involved some 6,500 people dealing in wholefood production and distribution (70 per cent); the remainder were in building, repairs, services, crafts and cultural/recreational activities.

## Portugal

One main example of LEI development is FNAC (National Manufacturing of Air Conditioning) which has grown into a major co-operative with 3,000 members. It provides free health services and housing allowance; its lowest paid members receive 40 per cent more than the average minimum wage. FNAC made a profit in 1983 of 150 million escudos which was divided as follows: investment expansion (70 per cent); reserve fund (5 per cent); education and training (10 per cent); social and welfare purposes (5 per cent); and salary-bonus schemes (10 per cent). The other co-operative, UNINORTE, is a new multi-purpose co-operative union mainly covering Northern Portugal. It provides legal advice, feasibility studies, education and training courses, trade fairs, etc. UNINORTE has a membership of 500 co-operatives comprising 5,000 members, of whom 2,000 have jobs in the co-operatives.

## Spain

Since 1984 Spain has had a specific employment programme entitled 'promotion of local initiatives for the creation of employment'. Its aim is to 'finance those initiatives which generate stable employment through the creation of small or medium-sized enterprises seeking to use unexploited resources in the district or region where they will operate and which have an innovative and stimulative effect on economic activity and employment'.

## United Kingdom

LEIs take four basic forms in the United Kingdom:

**Single-corporation-started projects** British Steel Corporation (Industry) Ltd, set up in 1975 as a wholly owned subsidiary of the BSC, has sponsored a number of interesting and unconventional job creation strategies at local level. These projects utilize and convert redundant premises, equipment, and facilities into starter workshops and mobilize direct or indirect financial support. It became active in 18 areas, where it nursed the establishment of small viable businesses. In Corby, where 140,000 jobs were lost, BSC helped to create 58,000 jobs in 2,300 local firms. Other large firms like United Biscuits, British American Tobacco, and lately the National Coal Board Enterprise Ltd are following BSC's example. Pilkington too is involved in this type of development and in venture capital experiments.

**Joint local enterprise initiatives** Between 1981 and 1986 the number of local enterprise agencies expanded from 23 to 250 throughout the UK. In 1985 48 per cent of their total cash income came from the private sector (6,000 sponsorships by over 3,000 companies), 27.5 per cent from local authorities, and 24.5 per cent from central government. Largely through their overall agency, Business in the Community, their co-operation with local authorities and collaboration with the Action Resource Centre, they have made the UK the prime leader in Europe in this type of local-enterprise-agency development. BIC claims that these agencies are now creating more than 50,000 jobs per year through business start-ups and securing the retention of 25,000 jobs annually through work with small firms.

**Community businesses** are set up and owned by the local community. They are multi-functional, providing local services and various schemes of community benefit; they aim primarily at creating jobs for local people.

**Local workers' co-operatives** Finally, LEI activity is exemplified in local workers' co-operatives. Their number rose from 234 in 1980 to over 900 in 1985, while the numbers of persons employed in these initiatives rose from 6,000 to 9,000 in the same period. Central government support for workers' co-operatives has been marginal.

## CONCLUSIONS: ASSESSMENT OF LEIs AND RECOMMENDATIONS

Local employment initiatives offer important opportunities in policy towards long-term unemployment. Their advantages can be summarized as follows. Local employment initiatives:

help to stimulate the creative and entrepreneurial talents of individuals at the local level;

give those who are involved in them a greater sense of local identity and a realization of the service that they can contribute to their community;

promote a sense of solidarity and collective effort as a desirable substitute for alienating and semi-adversarial worker/management relationships which sometimes characterize traditional forms of ownership;

provide a non-profit-orientated, enterprising, and productive form of self-expression and may often help to 'retrieve' persons who may be living on the margins of society and who run the risk of being completely alienated;

usefully combine professional and managerial training with sheltered market conditions;

transform distressed communities into 'opportunity' areas;

stabilize, involve, and integrate disadvantaged or marginal groups in the local community;

improve and promote local public facilities, services, products, and the environment.

At the same time there are many factors inhibiting the development of LEIs. These factors include:

the absence in some member states of the European Community of a sufficiently defined and flexible legal framework to promote different forms of LEIs;

specific legal difficulties in some member states for
new groups of self-employed people who are not
professionally qualified or recognized;

difficulties in other member states where people
wishing to form a worker co-operative are either not
legally catered for or classified as being self-employed
and therefore excluded from unemployment and
sickness insurance cover;

difficulties for workers' takeovers of bankrupt
enterprises, insolvency practice often not involving the
employees concerned and being usually dominated by
the legal issue of how to mediate between debtors and
creditors rather than how to rehabilitate the firm in a
new form;

particular legal problems for 'alternative' projects in
their 'non-ownership', 'neutral capital' structure; there
are difficulties in terms of recognition, registration,
and liability;

national or local political climates that may not at
times be conducive to creating the partnership essential
to starting up new enterprises;

over-centralized bureaucratic demands;

lower standards of pay, sometimes working conditions
which are not too good, and unsatisfactory insurance
cover and job security in LEIs;

a sometimes short lifespan;

an occasional antipathy towards trade unions;

the high degree of self-exploitation in 'alternative'
undertakings and the possible danger of accentuating
the 'black' economy;

the risk of training standards not being properly
monitored;

the feeling of many people, especially those with
immediate family obligations and responsibilities, that
they cannot risk short-term or dwindling personal
financial resources as collateral in an experiment in LEI
activity;

difficulties in arranging and financing feasibility
studies;

the rigid position of banks and public financial
institutions at local and national levels with regard to
taking risks in the direct financing of specific LEI
projects;

the difficulty of relevant and clear information getting
through to the localities on how to apply for and gain

access to relevant national and EC funds.

Against this background the Economic and Social Committee of the EC has remained an ardent champion of LEIs and of a positive role for the EC in their encouragement. In particular, it has produced the following recommendations:

Member states ought to calculate and provide the European Community with reliable data on the net job-creation impact of LEIs and how more sustainable jobs might be promoted in local economies.

Member states should not frustrate but instead encourage local groups, communities, and regions to secure an appropriate participation in decision-making with respect to the distribution and use of the EC funds available for the economic and social regeneration of their own areas, notably through the development of LEIs. This proposal applies especially to the more appropriate functioning and financing of the European Social and Regional Funds.

Up-front financing of viable LEIs could be facilitated by the use of substantial single grants paid out at the start of a project (for professional feasibility studies, etc.).

The European Community ought to promote the setting-up of professional advisory services for LEI development and build on the networking role of ELISE.

All persons employed or self-employed in LEIs ought to have access to state support in the form of unemployment benefit and other social insurance benefits, both in cases of failure and of temporary inactivity.

Workers' takeovers of insolvent or bankrupt enterprises could be facilitated by formally involving them in the liquidation procedures and by preventing an (uneconomic) shutdown of a factory by means of a jointly agreed care-and-maintenance contract procedure or through appropriate legal procedures in court, the interests of all parties concerned being taken into account with prime consideration being given to maintaining viable jobs.

Perhaps the judgement of the social partners in the Economic and Social Committee is best summarized in the

149

words of the Opinion of December 1986:

> The Committee recognizes the useful and at times vital role which LEIs have to offer, particularly in local and regional areas with acute unemployment problems. From a local or micro-economic point of view, LEIs can generally be identified as being advantageous and socially desirable. From a macro-economic point of view, LEIs cannot be seen as a panacea to mass unemployment. Their aggregate job-creative impact remains unclear. (12)

## REFERENCES

1   Commission of the European Communities (1987a) Memorandum from the Commission on Action to Combat Long-Term Unemployment, COM (87) 231 final, 18 May.
2   Commission of the European Communities (1983) Community Action to Combat Unemployment: The Contribution of Local Employment Initiatives, COM (83) 662 final, 21 November.
3   Economic and Social Committee (1984) Opinion on the Commission's Communication to the Council 'Community Action to Combat Unemployment' (529/84); also (1986a) Opinion of the Economic and Social Committee on Local Employment Initiatives (1061/86).
4   Commission of the European Communities (1983), ibid.
5   Organization for Economic Co-operation and Development (OECD) (1982) Co-operative Action Programme on Local Employment Initiatives, Paris: OECD, 10 October.
6   Commission of the European Communities (1983), p.9.
7   Commission of the European Communities (1987b) The New Technologies and Local Development, FAST series, no. 21, Luxemburg.
8   ibid., p. 137.
9   Holland, A., J. Daviter, and V. Gessner (1984) Legal, Fiscal, Social and Administrative Obstacles to the Development of Local Employment Initiatives, Bremen: Zentrum für Europaische Rechtspolitik.
10  Economic and Social Committee (1986b) Report of the Section for Social, Family, Educational and Cultural Affairs on Local Employment Initiatives (323/86).

11   ibid., p.7.
12   Economic and Social Committee (1986a).

# Chapter Eight

## LONG-TERM UNEMPLOYMENT AND LOCALLY-BASED INITIATIVES: WEST GERMAN EXPERIENCE

Manfred Kaiser

In Germany the first major measures of labour-market policy were taken in the course of the sweeping industrialization during the second half of the nineteenth century. At that time many workers were reduced to poverty simply because they did not know how and where to find appropriate work. Commercial employment agencies often exploited the plight of those in search of work by charging excessive fees. Other employment agencies, whose services were free of charge and which were organized mainly by charitable institutions, could operate only on a limited scale.

Before the turn of the century a few public local labour exchanges were set up, in particular in large towns, where difficult labour problems had first arisen. These exchanges did not, however, have an overall view of those larger sectors of the labour market, which were not confined to municipal boundaries, but were linked to the national economy and macro-economic factors. Endeavours to institute a comprehensive organization of the labour market received a new impetus, when, for the first time, mass unemployment occurred after the First World War.

Although Germany had taken the lead with her advanced social legislation among the large industrial countries of the world, the principle of unemployment insurance could not be realized until much later. Before the First World War the view was still prevalent that anybody who could not find work had only himself to blame. The thought that this situation could have anything to do with the state and development of the labour market appeared absurd. In so far as any aid was granted to the unemployed, usually by local authorities, it was in the form of poor relief.

The next step in the development of labour-market policy was to combine the placement service and unemployment insurance legally in one institution. The basic principle was that the provision of work should take priority over financial aid to the unemployed. With the Placement and Unemployment Insurance Act (AVAVG) of 1927 the National Office for Placement and Unemployment Insurance came into operation as a self-governing authority responsible for the administration of placement services and compulsory unemployment insurance. At this time unemployment in Germany grew dramatically. In July 1927 about 750,000 people were unemployed; by the end of 1928 the unemployment figure had reached two million. In 1932 registrations at the employment offices reached the highest figure, over six million jobless - almost a third of the working population. In those difficult days of economic depression, the main function of the National Office, that of providing jobs, had to be subordinated to the necessity of supporting the millions of unemployed.

At the beginning of the National Socialist regime in 1933 the National Office lost its dedicated function and its importance. Immediately after the Second World War the Local and Regional Employment Offices resumed their activities; they came under the control of the Ministers of Labour of the newly created federal states. Steps were soon initiated to transfer the placement service and the compulsory unemployment insurance for the whole federal territory, including West Berlin, once more to a self-governing institution. The Act of 1952 on the establishment of a Federal Office for Placement and Unemployment Insurance stipulated equal representation of the employers, employees, and public authorities, since the functions of the Federal Institute for Employment (Bundesanstalt für Arbeit) extend far beyond those of a mere insurance organization. Apart from unemployment insurance its main functions are a public placement service and vocational counselling.

The Placement and Unemployment Insurance Act of 1927 was replaced by the Employment Promotion Act of 1969. Since then priority has been given to the provision of a qualitative and quantitative balance of supply and demand in the labour market rather than to making compensation for loss and other activities. Under the provisions of the Employment Promotion Act, the Federal Institute for

Employment has, within the scope of the social and economic policy of the federal government, the special function of helping to ensure that neither unemployment, underemployment, nor a lack of manpower occurs or continues.

Inter alia the Federal Institute for Employment is obliged to inform the public about the state of the labour force and the labour market and to explain the situation and the developments in the labour market. It is compelled - even by requests independent of the placement services - to inform and individually advise employees and employers about the state and the development of the labour market, the trends in trades and professions, the necessity and possibilities of vocational training and training incentives, the incentives for entering into and retiring from employment, and about questions relating to the choice or filling of jobs. In this context the Institute for Employment Research (Institut für Arbeitsmarkt- und Berufsforschung), a department of the Federal Institute for Employment, analyses the past and future development of the labour market. Since the oil crisis of 1973 unemployment has increased, and labour-market projections indicate a continued high unemployment level for the Federal Republic of Germany in the 1990s (Klauder, Schnur, and Thon 1985; IAB-Kurzbericht 1987).

What are the prospects for the 1990s? Assuming that past trends in the labour-force participation rates continue, and that there will be a foreign migration balance of zero, the labour force will be reduced by 1.7 million between 1990 and 2000. The reason is a marked decrease in the number of births.

Even an average rate of economic growth of between 1 and 3 per cent per annum, combined with a reduction of working time of between 1 and 1.3 per cent for the supply of jobs, will not be powerful enough to balance the demand. So the employment policy of the federal government is under severe challenge to combat unemployment over the next decade.

## UNEMPLOYMENT AND LONG-TERM UNEMPLOYMENT: THE SITUATION IN WEST GERMANY

### Definitions, methodology, and data sources

For the purpose of labour-market statistics a person is defined as unemployed if he or she does not have a job; is looking for a job and is available for the placement services; and is registered as unemployed with one of the labour offices. A person is additionally characterized as <u>long-term unemployed</u> if he or she is permanently without a job for one year or more.

In the Federal Republic of Germany monthly unemployment surveys of the labour offices are supplemented by so-called 'structural surveys' of the stock of unemployed at least once a year, usually in September. The first surveys of this kind were conducted in 1973. Regular statistics on unemployment inflows and outflows were introduced in 1981. These flow statistics cover a period of two weeks in May/June each year, and they are organized as samples in such a way that their results represent estimations for a year as a whole (Brinkmann 1987). For international comparisons the annual labour-force sample surveys must also be mentioned. As unemployment in these statistics is not defined in the above way, the results are not taken into consideration in this report.

With respect to the length of unemployment the various statistics provide different findings. Stock statistics describe only a part of unemployment, namely duration, the <u>current</u> length of unemployment. They do not indicate the total length of unemployment and how it has been terminated. Only flow statistics are able to survey the <u>total</u> duration of unemployment and the reasons for leaving the register. Statistics of outflows from unemployment give answers to the following questions. Has a jobless person taken up a new job? Has he or she retired, or has he or she taken part in a training or retraining scheme? Has he or she found a job by personal initiative or with the aid of the placement services? How many placement efforts were necessary in order to find a new job? Many more questions can be answered by flow statistics. The results can be combined into indices, coefficients, and/or rates. Thus, a <u>re-employment rate</u> can be defined as the percentage of all employed against the total of unemployment outflows (Karr 1982).

It is not, therefore, surprising that the results derived from stock and flow statistics differ about length of unemployment. In 1986 there were 32 per cent unemployed for longer than one year according to stock statistics. In the outflow statistics, however, only 14 per cent fell into this category. A comparison between the findings of the two kinds of statistics indicates that the long-term unemployed are over-represented in the stock statistics; and that it is much more difficult for the long-term unemployed to change their status and to find new destinations than for the short-term jobless (table 8.1).

**Table 8.1** Duration of unemployment in stock and outflow statistics, September 1986

| Duration of unemployment | Stock of unemployed absolute | % | Outflows of unemployed absolute | % |
|---|---|---|---|---|
| less than 1 month | 248,618 | 12.2 | 15,025 | 16.1 |
| 1-3 months | 409,391 | 20.1 | 22,439 | 24.0 |
| 3-6 months | 320,037 | 15.6 | 23,412 | 25.0 |
| 6-12 months | 413,783 | 20.2 | 19,608 | 20.9 |
| 1-2 years | 333,992 | 16.3 | 8,926 | 9.5 |
| 2 years and longer | 320,016 | 15.6 | 4,142 | 4.4 |
| Total | 2,045,837 | 100.0 | 93,552 | 99.9 |

Source: Institute of Employment (1987), Unemployment Statistics, Amtliche Nachrichten der Bundesanstalt für Arbeit Sonderdruck: Arbeitsmarktanalyse 1986 anhand ausgewählter Bestands- und Bewegungsdaten, Nürnberg, pp. 642-3

## Unemployment: stocks, flows and developments

At the beginning of 1986 about 2.35 million people were registered as unemployed in West Germany; this figure decreased to 2.22 million at the end of 1986. More than 3.6 million persons became newly jobless, and about 3.8 million ended their unemployment status during 1986. More than 70 per cent of the latter or 2.5 million found new jobs. The number of unemployed averaged 2,228,000 in 1986, 76,000 or 3.3 per cent less than in 1985.

The increase of unemployment in the 1970s and early 1980s was accompanied by a rapid growth in the length of unemployment. In 1977 the number of long-term unemployed was about 130,000 or 14.3 per cent of the total unemployed; in 1982 this number grew to 327,000 or 18 per cent of the total, and in 1986 654,000 or 32 per cent had been jobless for a year and longer. Three phases can be distinguished in the Federal Republic: the first phase lasted up to 1980, with about 112,000 long-term unemployed on average per annum; the second phase covers the period 1981 to 1984, when a rapid increase of long-term unemployment from 163,000 to 617,000 took place; and during the third phase, since 1985, long-term unemployment stabilized at a high level between 650,000 to 670,000 people. A breakdown by duration of unemployment indicated at the end of September 1986 is shown in table 8.2.

**Table 8.2** A breakdown by duration of employment at the end of September 1986

| Duration of unemployment | Absolute | Percentage of total unemployed |
|---|---|---|
| 1-2 years | 334,000 | 16.3 |
| 2-3 years | 143,000 | 7.0 |
| 3-4 years | 81,000 | 4.0 |
| 4 years and longer | 96,000 | 4.7 |
| Total of long-term unemployed | 654,000 | 32.0 |

Partly due to the increase in the duration of unemployment and partly due to legislative changes, the proportion of the unemployed receiving unemployment benefits fell below 70 per cent (per annum average). In particular, the proportion of those receiving unemployment benefits (Arbeitslosengeld) decreased sharply, with some increase of the numbers of those receiving supplementary benefits (Arbeitslosenhilfe) (Brinkmann 1987). There is

**Table 8.3** Unemployed by kind of qualification and duration of unemployment in 1986

| | All unemployed | Long-term unemployed (more than 1 year) | | |
|---|---|---|---|---|
| | | Total | Male | Female |
| | % | % | % | % |
| Total. | 2,045,837 | 654,008 | 336,038 | 317,970 |
| % | 100.0 | 100.0 | 100.0 | 100.0 |
| | | | | |
| No formal vocational training qualification | 50.8 | 56.7 | 55.5 | 58.1 |
| No graduation from primary school | 15.1 | 17.3 | 20.9 | 13.5 |
| Primary-school graduate | 35.8 | 39.5 | 34.6 | 44.6 |
| Level of a 'vocational training school' | 49.2 | 43.3 | 44.5 | 41.9 |
| Apprenticeship | 38.7 | 34.9 | 37.4 | 32.4 |
| Vocational training* | 5.0 | 3.7 | 2.5 | 5.0 |
| Fachhochschule (college) | 1.7 | 1.3 | 1.4 | 1.1 |
| University | 3.9 | 3.3 | 3.3 | 3.4 |

* Berufsfachschule and Fachschule

Source: Institute of Employment, Unemployment Statistics

evidence that the increase of long-term unemployment is correlated with a decline of unemployment benefits.

## Some characteristics of long-term unemployment in West Germany

The increase of long-term unemployment stems from the dynamics of the labour market (Büchtemann 1984). It reflects risks connected with unemployment, notably selection processes which may be dominant in some years and not so much in others, including:

people with health problems, disabled and handicapped.

people in higher age groups, so that the proportion of long-term unemployed (among the total of unemployed) increases in higher age groups. A breakdown of all unemployed compared with long-term unemployed reveals that about 38 per cent of the total are older than 40 years of age but 56 per cent of the long-term unemployed belong to this age group; 22 per cent of the long-term unemployed are under 30.

first-time jobseekers (young people) and new entrants into the labour market.

females, particularly because of their lower qualification status, their reduced availability for job options, their wish for part-time jobs, their (recurrent) interruptions of employment, frequently for family reasons, and their increasing participation in the labour market.

people with insufficient vocational qualifications. Thus, about 51 per cent of all unemployed did not acquire a formal vocational qualification; the percentage among the long-term unemployed amounted to about 57 per cent (male: 56 per cent; female: 58 per cent) (table 8.3). The difficulties become extremely great, if the status of an unskilled or semi-skilled person is combined with an age of 55 years and over and/or with health problems. This problem has been confirmed by an analysis of outflows of unemployment for the year 1986 (see table 8.4). One out of seven (14.1

**Table 8.4** Outflows of unemployment by duration of unemployment and special characteristic/characteristic-combinations, 1986

| Characteristics/ characteristic- combinations | Outflows of unemployment | Duration of unemployment | | | sub-total | | Number of unemployment cases |
|---|---|---|---|---|---|---|---|
| | | 1 to 2 years | 2 years and longer | 1 year and longer | Up to 1 year | |
| | % | % | % | % | % | % | |
| NG + HP + older 55 | 100.0 | 22.1 | 27.4 | 49.5 | 50.5 | 551 |
| NG + HP | 100.0 | 14.8 | 11.8 | 26.6 | 73.4 | 4,290 |
| NG + older 55 | 100.0 | 22.3 | 16.9 | 39.2 | 60.8 | 1,155 |
| HP + older 55 | 100.0 | 29.2 | 25.3 | 54.5 | 45.5 | 538 |
| NG | 100.0 | 10.1 | 4.6 | 14.7 | 85.3 | 32,415 |
| HP | 100.0 | 12.3 | 6.5 | 18.8 | 81.2 | 4,691 |
| older 55 | 100.0 | 25.2 | 16.5 | 41.7 | 58.3 | 1,060 |
| none of these characteristics | 100.0 | 7.6 | 2.5 | 10.1 | 89.9 | 48,852 |
| Total 1986 | 100.0 | 9.6 | 4.4 | 14.1 | 85.9 | 93,552 |
| 1985 | 100.0 | 12.0 | 3.9 | 15.9 | 84.1 | 104,928 |

Characteristics    NG: no formal vocational-training
                      HP: health problems
                      older 55: older than 55 years of age

Source: Institute of Employment, Unemployment Statistics
Brinkmann, C. (1987) Langzeitarbeitslosigkeit – Stand, Entwicklung, Perspektiven, Massnahmen, Nürnberg

per cent) was jobless longer than one year on the average. However, if they did not acquire a vocational training grade, they were older than 55 years, and they additionally suffered from health problems, one in two of them (49.5 per cent) belonged to the group of the long-term unemployed. The most risky unemployment components are health problems and a relatively high age.

## Transition from long-term unemployment into working life (re-employment)

Efficient and effective employment policies depend on valid and reliable answers to the following questions. What are the careers of the long-term unemployed? In what way and through which channels do or can they end their unemployment status? What are the main factors disturbing or hindering their placement and re-employment opportunities? The following results from a sample survey of unemployment outflows in spring 1986 show important details about the transition from unemployment into a new destination.

There were about 95,000 unemployed who ended their unemployment status. Seventy-three per cent of all the outflow unemployed (= re-employment rate) found work (28 per cent with the aid of the placement services; 40 per cent by their own initiative; and 4 per cent by moving into job-creation measures); 1 per cent participated in special measures organized by companies, and about 4 per cent participated in continuous training/retraining and rehabilitation measures. The destination of the rest is unknown, but it must be assumed that most of them have stayed at home or receive pensions. The new Employment Promotion Act has made it possible for older employees (older than 58 years of age) to leave employment and receive unemployment benefits until reaching regular retirement age.

Fourteen per cent (14,000) of the group of outflows (95,000) were longer than one year without work. Their re-employment rate reached 52 per cent: 18 per cent found jobs with the aid of the placement services; 23

per cent found work through their own initiative; and 9 per cent were placed by job-creation measures. Two per cent participated in special measures organized by companies, and another 6 per cent have been continuously trained or retrained or participated in rehabilitation measures. The status of about 42 per cent is, however, unknown.

About 4.5 per cent (4,160) of the outflows were longer than two years without work. Their re-employment rate amounted to about 49 per cent. Their employment profile is similar to that of the group of outflows who were unemployed longer than one year, with the exception of their participation in job-creation and training schemes, which was higher.

## Some conclusions

Long-term unemployment is a particular aspect of unemployment and reflects the way industrialized economies and labour markets operate. It results from a number of factors.

The inadequate rate of new job creation, though the importance of job-creation programmes has grown for the long-term unemployed.

The industrial and regional concentration of the effects of structural and technological change over the past decade demands training and retraining requirements to an extent not required in the past.

The way in which the labour market works for those unfortunate enough to lose their jobs and who bear most of the burdens of adjustment. The victims of these adaptation processes are mainly the unskilled and semi-skilled, the disabled and handicapped, and the older employees. Furthermore, long-term unemployment has detrimental effects on health, mainly in relation to mental well-being, but also concerning physical health. The chicken-or-egg situation arises of which came first unemployment or health problems?

The fact that unemployment and social securit

systems are not necessarily geared to giving the right support or incentives for the re-employment of the long-term unemployed (Commission of the European Communities 1987, p. 17).

Since the mid-1970s, new types of employment have developed in West Germany as a result of unemployment, continuing underemployment, and a change in individual and social attitudes towards work. It is not yet possible to determine what importance these new forms of employment will have in the future, whether they are permanent, and whether they are voluntary or forced. A 'second' or 'additional' labour market is emerging in West Germany. This 'market' has increased during the last few years and is still growing. Originally, it was characterized mainly by job-creation schemes (Arbeitsbeschaffungsmassnahmen or ABM). Now it is supplemented by further welfare measures in the fields of vocational training and employment (e.g. social assistance, supplementary benefits, child and youth welfare). It is aimed at (re-)integrating unemployed people who have no (further) claims according to the Employment Promotion Act (Arbeitsförderungsgesetz, AFG). The 'additional labour' market is a subsidized job market to a large extent. The subsidies are paid mainly by public organizations, namely the Federal Institute for Employment (Bundesanstalt für Arbeit), the federal states, and local authorities. Churches, welfare organizations, and local employment initiatives are often the agents of these schemes.

## LOCALLY-BASED INITIATIVES IN WEST GERMANY

In the Federal Republic of Germany four important developments in the area of locally based initiatives can be identified: unemployment initiatives; 'alternative-economic' initiatives; social work and employment initiatives (or social initiatives); and counselling initiatives, aimed primarily at supporting the above-named initiatives with, for instance, welfare services. The general term, 'local employment initiatives', summarizes the various projects and self-help groups, both private and public, which have a goal of creating 'meaningful jobs' and which deviate from traditional firms and institutions in their organization and/or in their manner of foundation.

163

## Unemployment initiatives

Self-help establishments for unemployed people are known under the names of 'meeting places for job seekers', 'unemployed initiatives', and 'centres for jobless'. Exact definitions of these terms have not yet materialized. Many of these institutions are staffed by full-time personnel trained in social work and psychology in order to facilitate co-operation both internally and externally. An initiative is more likely to receive financial security through public promotion if it is tied to a public agency.

In 1983/4 it was estimated that there was a total of 300 to 500 unemployment initiatives in the Federal Republic. Due to the rise in unemployment, there has been a dramatic increase in the foundation of new unemployment initiatives. Since the beginning of the 1980s, for example, the state of North-Rhine Westphalia has been covered with a network of unemployment initiatives; most of them are concentrated in densely populated areas and in major cities. In spring 1983 there were only 70 unemployment initiatives there; by autumn 1986 the figure had risen to 260.

Few unemployment initiatives (probably 10 to 20 per cent) have workshops and training facilities at their disposal, run by older, unemployed craftsmen. As far as the total number of available training positions is concerned, the training possibilities in the unemployment initiatives play only a marginal role. Very often, young people leave the training position as soon as they are offered a regular position in a firm/factory. These initiatives concentrate on those unemployed persons who are difficult to place, i.e. those with fewer capabilities, the handicapped, maladjusted, un- and semi-skilled teenagers, and young adults. Mainly teenagers and young adults up to 25 years of age are approached, helped, and partially qualified.

The future importance of the unemployment initiatives and centres is dependent upon: whether, and to what extent, unemployment - particularly long-term unemployment - in Germany spreads; which social groups will be affected; and what kind of relationships the unemployment initiatives will develop to the existing large organizations, i.e. trade unions, employers' associations, governments, and local authorities.

The unemployment initiatives have developed the following goals and tasks:

removal of isolationist tendencies;

promotion of the feeling of self-esteem and a realistic self-evaluation;

maintenance and development of important basic skills, e.g. of communication, co-operation, and decision-making capabilities;

acting as co-ordinator between the various authorities and institutions (employment offices, youth welfare offices, churches, trade unions, politicians, employers, etc.);

representing the interests of unemployed people to the public and within firms, carrying out relevant public-relations campaigns;

organizing and diffusing ideas, particularly relevant to planning the daily life of unemployed people with respect to recreational and educational possibilities;

maintenance of the relationship to the working and professional worlds by co-operating with the institutions dealing with the labour market and by meetings with employed people with the goal of maintaining placement possibilities ('working capability cannot be stored');

research, testing, and utilization of employment opportunities, particularly in the 'additional labour market' (including job-creation measures);

socio-political work, for example, aimed at local social improvements (e.g. reduced public-transport fares or unemployment identity cards).

In times of growing polarization of unemployment, such unemployment initiatives are particularly important for maintaining social peace.

## 'Alternative-economic' initiatives

The major German specialist on alternative movements, U. Mäder (1980), sees the 'alternative-economic organizations', and the 'newly self-employed', as an indication that there are people who are no longer willing to:

work in organizations, where there is a clear boss-subordinate relationship;

function as a number or part in a large, impersonal, complex company;

produce goods for which they can assume no responsibility;

withstand stress on the job;

give up the direct contact with customers for whom they are actually producing;

separate their political activities from their jobs, where they earn their living;

withstand unemployment or low-value, senseless, and alienating work.

The new formula in the 'alternative economy' or 'self-governed' economy is cost recovery instead of profit planning, co-operation and networking instead of competition, equal rights instead of hierarchy, job rotation instead of job monotony, capital neutralization instead of individual ownership, uniform or need-oriented wages instead of wage differentiation, and democratic decision-making instead of individual decision-making. Thus the basic characteristics of the self-governed, alternative economy are production in organizational units which are understandable to all concerned, participation in decision-making by all members of the respective production units the realization of a rotation principle for all activities/jobs an equal disbursement of the jointly produced profit, the abolition of so-called wage labour, and capital neutral ization, with the goal of making social changes practical through self-organized forms of work and co-existence.

The size of the entire alternative-economic sector i estimated at 11,000 to 13,000 projects, engaging 80,000 t 130,000 people, with a statistical average of approximatel seven members per project. Taking into account empirica findings that one out of four persons will be fully paid fo his job, guaranteeing subsistence, the entire work force in the alternative-economic sector can be estimated at between 25,000 and 40,000 economically active people (Kaiser 1985).

The alternative-economic projects are characterized by a high labour intensity. In general, 70 per cent of the alternative-economic projects can be classified in the service sector, in the broadest sense of the term, within which welfare-service jobs are the most significant with 22 per cent (school projects, care for children and the elderly, medical groups, therapeutic, social work and youth welfare projects, centres for women, etc.). The area of small production covers between 12 and 25 per cent of the jobs (e.g. print, production, and repair shops, alternative technological organizations, handicrafts).

Self-governed projects employ mainly men and women between 20 and 30 years of age with an above average education (Hegner and Schlegelmilch 1983, p. 37).

About 40 per cent of the employees in the projects have graduated at a university or are still studying (in comparison, the working population contains only 8-9 per cent university-educated people); two-thirds of these are social scientists, social workers, and teachers and are often working in non-professional positions in the projects. These latter groups often have a high unemployment rate and long duration of unemployment.

The number of people who stated that they had just 'some kind of job' before beginning their position in the alternative-economic project is unusually large.

About 16 per cent of the employed are trainees.

Account has to be taken of the frequent divergences between a person's educational status, the job to be performed, and the ability and need to carry out managerial functions. The co-ordination or adaptation of these three functions sets a multitude of learning processes into action in the self-governed projects and is necessary if the self-governed project wants to survive economically. The learning processes in the projects are focused on the acquisition and use of specialized knowledge and capabilities. These learning processes result from the fact that the education of the employees and the job areas in

167

which they work are very often incompatible. Thus specialized knowledge must first be intensified or newly acquired and then put to use. The learning processes emphasize manual skills, personality-related characteristics, and social competence.

Clearly these new types of job have managerial characteristics, such as aspiration for independence, acceptance of risk and responsibility, resource allocation and work planning, working at one's own cost, and production of goods and services which ensure one's existence. The catchword 'new self-employed' has therefore a real basis. It can be seen that the alternative projects develop job profiles which more strongly resemble the activities of a freelance than those of an employed person.

## Social work and employment initiatives (social-employment initiatives)

In addition to unemployment initiatives, alternative-economic projects, and the promotion of self-help groups, social-employment approaches and strategies are becoming more and more part of the discussion and practice of an experimental employment policy (Kaiser 1987). If they also offer counselling and care services to special groups of people, as an independent legal entity with a welfare character, then they are classified under the general term 'social work employment initiatives' (or social employment initiatives). These initiatives can be recognized by the following characteristics: specialized goals, specific job categories, and the social content of their activities.

The specific target groups of social-employment initiatives are primarily socially disadvantaged, unemployed people and/or difficult-to-place people, according to the Employment Promotion Act (AFG) and to youth and welfare legislation. These groups are subject to extreme competition in the fight for employment. To a large extent, they are rejected by the labour market due to their lack of qualifications, their lack of professional experience, their often prematurely ended training and/or professional education courses, their outdated professional education or job experience, and finally their health, psychological and social problems. Many social-employment initiatives have, therefore, based their programmes on personal, educational, and practical qualifications in order to improve the work and professional prospects of their target groups.

Second, the majority of the social employment initiatives are geared towards the service sector, in the broadest sense of the term. They carry out mainly non-profit-making work, aimed at the public interest. The activities are often limited in duration and have a relatively limited scope. The areas of environmental protection, improvement of living environment, culture, energy-saving, measures for improving living standards, as well as the production of social necessities and additional goods and services, are all seen as excellent fields of activity for social-employment initiatives.

Third, social-employment initiatives fulfil, among other things, the following criteria:

they seek integration or re-integration into the social and value system;

they seek to promote activity within the wider economic system;

educational qualifications are obtained in recognized professions;

those employed usually fall within the scope of the social-insurance system;

they possess an employee status;

remuneration is usually based on official wage rates; and they receive wage subsidies from public funds (e.g. the Employment Promotion Act (AFG), Federal Welfare Act (Bundessozialhilfegesetz), Child and Youth Welfare Act (Jugendwohlfahrtsgesetz), training/retraining and continuous education programme within the framework of the Employment Promotion Act, and the federal programme for the disadvantaged).

A distinction needs to be made between social-employment initiatives in a broader and in a more limited sense. The distinguishing factor is the responsible agency. In so far as initiatives have been formed as individual legal entities, which in addition fulfil the three criteria mentioned above, then they are considered as social-employment initiatives in the strictest sense. It is of no significance whether an initiative was formed with an educational or employment goal, or whether it was oriented towards a specific problem which then leads to educational and/or employment effects. By contrast, if educational and employment goals are assumed by traditional organizations, clubs and associations in addition to their other goals or responsibilities (e.g. by churches, welfare organizations, local authorities), then these activities are not considered to be 'educational or employment initiatives' as such, but rather social initiatives in the broadest sense of the term.

In the following estimate, only social-employment initiatives in the strictest sense are considered. It refers to the expansion of such employment initiatives in 15 employment office districts in the Federal Republic of Germany which participated in the job-creation scheme (Arbeitsförderungsmassnahme, ABM) in November 1986. These 15 districts are representative for the Federal Republic as a whole. Between 2,400 and 3,200 social initiatives with a total of 60,000 to 80,000 members were identified. This figure includes between 41,600 and 55,600 members who received an income sufficient to ensure their subsistence level or at least additional subsistence funds. A basic prerequisite for the creation and effectiveness of social initiatives is the co-operation of non-paid members, persons who work regularly in the initiatives without an income. The work carried out by non-paid members often forms the basis for the establishment and development of the project and it is also imperative if the full-time employees are to achieve successful and innovative work.

Various opportunities exist for funding social initiatives:

under the Employment Protection Act (Arbeitsförder-ungsgesetz, AFG), e.g. financing job-creation schemes, based on Sections 91-6 of the Act and other AFG provisions (e.g. for retraining, further education, on-the-job training, tideover allowance);

under the Federal Welfare Act (Bundessozialhilfege-setz, BSHG), e.g. creating jobs for unemployed welfare recipients who do not receive payments according to the AFG, based on Section 19, para. 1 BSHG;

under the Youth Welfare Act (Sections 5, 6 and 62 ff Jugendwohlfahrtsgesetz, JWG) and promotiona schemes for child and youth welfare (e.g. federal youth plan, German youth stamp foundation);

under the federal programme for the disadvantaged;

under the Severely Disabled Persons Act (Schwerbehindertengesetz), for the promotion of measures to integrate severely disabled persons into the labour market and society;

state (Land) policies, under which certain measures are promoted (see scheme in North-Rhine Westphalia 'Work Rather than Welfare');

local policies which are used together with the state measures in order to fulfil certain regional priorities;

through the use of conscientious objectors and trainees (e.g. within the framework of training courses offered by specialized colleges of higher education);

funds of associations and agencies that they have mainly earned by themselves (e.g. ground dues from a kindergarten playground, care allowance for the elderly and from donations, fines, association dues, funds from the child welfare campaign or youth stamp campaign, funds from public agencies). It is difficult to link the number of paid employees to the main sources of financing because the contributions to the initiatives are often not paid according to material, personal, or project-related criteria. In particular, the category 'agencies/associations' own funds' implies that the initiatives actually have their own funds, and that is rarely the case. Their resources come usually from various external sources, differing from case to case, and involving different combinations of financing.

The social initiatives finance their paid employees to the extent of approximately 46 per cent from their own funds; 25 per cent from ABM funds; 1 per cent from other AFG funds; 15 per cent from federal welfare funds; 5 per cent from the child and youth welfare programmes; 6 per cent through conscientious objectors; and 2 per cent through trainees.

The character of all promotion measures and financing

schemes is based on the regulatory and socio-political goal of integrating the jobless into the general labour market. In this respect, certain criteria are usually applied to measures and to the specific demands on the claimants: the duration of the measure is limited; the claimants are allocated for a limited time; and the circle of claimants is legally controlled and quantitatively limited. Many social projects and initiatives could not exist or would never have come into existence without the support of job-creation schemes.

Within the social initiatives themselves one must differentiate between the project management, instructors, welfare personnel, and the persons to be served and helped. For the managerial duties in particular, a qualified staff is required. Academically educated or other highly qualified unemployed people have benefited disproportionately from the job-creation possibilities. For example, about 38 per cent of the members of social initiatives have a university degree, another 12 per cent have completed an upper secondary leaving certificate, and approximately 50 per cent have completed a lower secondary leaving certificate or have an intermediate leaving certificate. Usually the college-educated and more highly qualified people are in a position to develop and carry out 'meaningful' projects within the framework of the available promotion possibilities. This phenomenon illustrates the innovative ability of university-educated people in phases of employment crisis. Using social initiatives as an example, we can also show that the implementation of social innovations benefits disadvantaged groups considerably. This benefit results from the fact that educated persons disadvantaged by the labour market, as well as social fringe groups, gather professional experience within the social initiatives, maintain or promote their ability to make social contacts, and achieve additional qualifications.

In principle, social initiatives can fulfil three different functions. Through their anti-cyclical effects the labour market can be temporarily relieved; through a problem-and group-oriented approach hard-to-place people can be (re-)integrated into the labour market; and by using experimental promotion measures, the creation of new jobs in the area of social needs can be strengthened.

## Counselling initiatives

Counselling initiatives do not directly benefit the unemployed. Their effects are indirect, in so far as they attempt to strengthen and upgrade other types of initiative and self-help groups by counselling and giving advice.

The political structure of the Federal Republic allows the states and local authorities their own quite extensive administrative freedom so that a complex evolution of various support agencies has developed (child and youth welfare, social assistance, advanced training agencies, adult education centres, etc.). Thus the numerous counselling establishments are based on federal, state, and local levels. More recently, however, the activities of agencies have begun to overlap with respect to the promotion of local training and employment initiatives. The following types of joint action have started to emerge: the combination of regional economic development programmes with regional labour-market strategies to promote locally-based initiatives; co-ordination of welfare measures under the youth and social assistance policies with those of the classic labour-market policies under the Employment Promotion Act; co-ordination of the independent initiatives of the alternative movement (networks) with the national promotion programmes; and attempts to co-ordinate various promotion and action measures through new systems of co-operation.

This development of joint action is understandable when one considers that, whilst the new initiatives (with regard to their type of organization, their duties and working methods) usually do not lack commitment and ideas, they have problems in understanding and utilizing existing promotional possibilities. In addition, the self-help initiatives often lack experience, especially in the area of economic organization, notably management and accounting. Due to the unconventional character of these initiatives, their lack of experience cannot be overcome with the help of established counselling services. Sometimes they cannot even be reached by such services.

The groups requesting counselling services can be divided as follows: cities (medium to large-sized cities, including the city states) and their districts; welfare associations and further training agencies; initiatives, trade unions, projects etc. The counselling services are focused on the following problem areas:

development of financing models and strategies;

development of training programmes, derived from the needs of specific initiatives and self-help groups;

dissemination of information about training schemes and methods for initial professional training;

development of new training jobs, as well as finding the necessary investment means;

development of appropriate organization and co-operation models for employment companies (Beschäftigungs GmbHs) and project groups;

application of management techniques for leading the projects and for programme management;

counselling services when submitting applications and accounting terms of agreement for financing under the European Social Fund (ESF).

Experience with this counselling approach in West Germany has revealed the following trends:

Despite a continued need for training and apprenticeship opportunities for certain youth groups and in certain regions, the projects and promotion efforts are increasingly aimed at combined training and employment measures. In addition to the effects of demographic development, this trend is due to the difficulty in placing trained young people in regular jobs. For this reason financing agencies are reluctant to promote new training capacity.

The training and employment measures follow two directions: relieving the labour market through the subsidized, long-term employment of older long-term unemployed people; and by the training of younger long-term unemployed people, reducing local disparities between job vacancies and qualified applicants.

When training and employment measures are combined with regional economic-assistance policies, there is a automatic demand for product and service innovations.

174

## CONCLUSIONS AND RECOMMENDATIONS

1   The jobs in the West German alternative economy and in social initiatives supplement existing activities in the private and the public sectors. Work in these new areas is often collective and lowly paid in permanent positions. At the same time it is often subjectively satisfying. This type of job is not reduced simply to a source of income; it is an important part of an individual life style which allows the development of one's personality (Beywl, Brombach, and Engelbert 1984).

2   The success or failure of local employment initiatives should not be assessed merely in terms of economic efficiency and effectiveness. In an overall evaluation, the following factors must also be considered:

professional re-employment and (re-)integration of unemployed persons;

reduction of the economic costs caused by unemployment;

removal of social tensions;

social benefits for the consumers of products or services;

regional development of purchasing power;

reduction of illicit work;

positive effects on the environment;

meaningful utilization of resources and recycling possibilities.

3   Even if local education and employment initiatives are dependent to a large extent on subsidies, these funds are used in such a way that they enable unemployed people to earn a living rather than simply to receive a subsistence allowance (unemployment or supplementary benefits, social assistance).

175

4    The initiatives' main problems are due to financing bottlenecks during the founding phase and to the specific make-up of the groups of people concerned. The participants are not always appropriately qualified with respect to their professional abilities or do not have sufficient know-how successfully to establish an employment record. Even if some employment initiatives cover 50 per cent of their costs themselves, and even if in the coming years greater efficiency is expected, these initiatives and projects will still depend on grants. This situation is particularly true for the social initiatives (Hegner and Schlegelmilch 1983).

5    Given the special character of the employment initiatives and the background of considerable difficulties in placing long-term unemployed people in work, the following consequences and requirements arise for securing jobs on a long-term basis:

(a)   By way of example, regional or local demand analyses must be made for occupations, professions, services, and products in order to ascertain existing and required training and employment potential, locally or regionally (by private companies, education and employment initiatives, alternative-economic projects, or the state).

(b)   Existing employment possibilities and new fields of employment must be increased in order to ensure a sufficient number of employment openings for unemployed people who will not be able to find a job on the labour market on a long-term basis.

(c)   Medium- and long-term training and employment strategies should be developed which enable a permanent return to the general labour market; these strategies should be based on local authorities or city districts.

(d)   Local and regional policy integration is necessary to agree upon the content, organization, and financing of initiatives, and for increasing the efficiency and development as well as diffusion of training measures. At the same time, this policy integration helps to avoid the distortion of competition, to promote an interrelationship of new employment approaches and local needs and to open new possibilities for marketing products and services.

(e) Local authorities and welfare organizations should delegate more tasks to the employment initiatives, which would result in an increase in contracts granted by the state.

(f) Counselling centres should be promoted, in so far as they represent the interests of local employment initiatives, and thus create an important precondition for the development and existence of local unemployment, education, and employment initiatives. It is particularly important to offer appropriate educational and further training courses to counsellers working in these centres.

(g) The additional specialized know-how of company or state specialists (e.g. regional planners, engineering specialists, technicians) should be integrated into the self-governed and social-employment initiatives.

(h) Social-oriented and environmental improvement jobs should be initiated and promoted as a priority.

(i) A number of measures and projects demand social-work orientation and guidance. The use of qualified training and welfare personnel with guaranteed financing should be promoted. Regular further education and training programmes should be guaranteed.

(j) The variety of financial promotion possibilities through the federation, states, local authorities, public employment service, etc. often leads to a lack of co-ordination and results in a limitation of the possibilities for long-term employment. Public means should therefore be more strongly interrelated with regard to collectively agreed regulations.

(k) The respective legal regulations as well as the federal and state programmes should contain measures for training and professional qualification in local employment initiatives, above all with regard to managerial capabilities.

(1)  The labour-market policy of the Employment Promotion Act should be better adapted to the changed conditions on the labour market. Eventually new measures of job creation must be developed and/or existing ones modified, including:

>    expansion of the number of people helped by locally-based initiatives;

>    extension of the duration of help in individual cases (i.e. more flexible management of training time, according to need);

>    in addition to helping people more direct support also for the projects;

>    guarantee of the availability of social-work care and counselling personnel;

>    more flexible definition of the criteria 'addition', 'supplement', 'public interest', and 'limitation of the job-creating measures' in approving the work to be carried out;

>    increase and redistribution of the means of assisting the long-term unemployed to the benefit of local training and employment initiatives.

## REFERENCES

Beywl, W., Brombach, H., and Engelbert, M. (1984) Alternative Betriebe in Nordrhein-Westfalen, Düsseldorf.

Brinkmann, Ch. (1987) 'Unemployment in the Federal Republic of Germany: recent empirical evidence', in P. Pedersen and R. Lund (eds.) Unemployment: Theory. Policy and Structure, Berlin/New York, pp. 285-304.

Brinkmann, Ch. (1987) Langzeitarbeitslosigkeit - Stand, Entwicklung, Perspektiven, Massnahmen, Nürnberg. Dortmund, unpublished.

Bundesanstalt für Arbeit (1987) 'Arbeitsmarktanalyse 1986 anhand ausgewählter Bestands- und Bewegungsdaten Ergebnisse und statistische Übersichten de Sonderuntersuchungen von Mai/Juni 1986 und End

September 1986' in Amtliche Nachrichten der Bundesanstalt für Arbeit, no. 3 und no. 5.

Commission of the European Communities (1987) 'Actions to combat long-term unemployment' in Social Europe, no. 2, pp. 17-23.

Hegner, F. and Schlegelmilch, C. (1983) Formen und Entwicklungschancen unkonventioneller Beschäftigungs-initiativen, Berlin, unpublished.

IAB-Kurzbericht vom 8.12.1987, Zur Arbeitsmarktentwicklung bis 2000, Nürnberg, unpublished.

Kaiser, M. (1985) 'Alternative-ökonomische Beschäftigungs-experimente - quantitative und qualitative Aspekte: Eine Zwischenbilanz' in Mitteilungen aus der Arbeitsmarkt- und Berufsforchung, no. 1, pp. 92-104.

Kaiser, M. (1987) 'Qualifizierung in Beschäftigungsinitiativen: Herausforderungen an eine lokale Bildungs- und Beschäftigungspolitik' in Mitteilungen aus der Arbeitsmarkt- und Berufsforschung, no. 3, pp. 305-20.

Kaiser, M. et al. (1987) 'Qualification and employment initiatives in the Federal Republic of Germany - ideas for local employment initiatives, Nürnberg, paper to be presented for OECD, unpublished.

Karr, W. (1987) 'Bewegungsgrössen in der Arbeitslosen-statistik' in D. Mertens (ed.) Konzepte der Arbeitsmarkt- und Berufsforschung (Beiträge zur Arbeitsmarkt- und Berufsforschung), Nürnberg, pp. 333-45.

Klauder, W., Schnur, P., and Thon, M. (1985) 'Arbeits-marktperspektiven der 80er und 90er Jahre: Neue Modellrechnungen für Potential und Bedarf an Arbeitskräften' in Mitteilungen aus der Arbeitsmarkt- und Berufsforschung, no. 1, pp. 41-62.

Mäder, W. (1980) 'Ich konnte es nie ertragen, unter einem Chef zu arbeiten. Alternativbetriebe - Endstation für Ausgeflippte oder zukunftsweisende Berufsperspektive' in Perspektiven, no. 4, p. 52f.

## Chapter Nine

## STRATEGIC RESPONSES OF LOCAL AUTHORITIES IN THE UNITED KINGDOM TO LONG-TERM UNEMPLOYMENT

Derek Portwood

Long-term unemployment in the United Kingdom is characterized by its persistence on an immense scale (1) and its uneven social and geographical distribution. (2) Strategic responses can concentrate on one of these features or on both. The choice of strategic response is partly determined by availability of resources and political priorities with respect to their allocation. It also hinges on political sensitivity about the popularity of central and local government, particularly the need to be seen to be caring, even-handed, and effective.

For central government the initial choice must be to tackle the magnitude of the problem of long-term unemployment. Obviously state income maintenance is the first priority. At the same time, tensions if not conflicts between central and local government (e.g. over housing benefit and rent arrears) make the purposes, rules, organization, and delivery of that provision highly complicated. In any case social-security arrangements have increased the levels of demand for local authority services from long-term unemployed people. (3)

The other pressing demand on central government is to find or create jobs for unemployed people or at least prepare them for entry (often this is re-entry) into work. In the United Kingdom this has led to a series of national programmes designed for all long-term unemployed people, mainly from the Training Commission (e.g. Restart). In addition, the Department of Education and Science has contributed to the development of educational opportunities for unemployed adults through its Replan programme. (4) Hence, while the objectives have differed (e.g. the

180

Community Programme is concerned with work experience and Restart with a variety of work, job searching, and training options), the provisions have been for all long-term unemployed people. Even the Job Training Scheme which originally was directed mainly at under-25s has now been combined with the Community Programme and made universally available to all the long-term unemployed by the Training Commission (now the Training Agency).

Local government necessarily starts from a different position. The framework of national provision is predefined. Whilst its own ideological preferences will shape its views and use of such provisions, this provision cannot be ignored if the scale of the problem is to be addressed. In any case, local government has its hands full as a consequence of the uneven distribution of long-term unemployment. Certain age, family, and ethnic-minority groups are heavily over-represented and escalate demands especially in the areas of health, housing, and social services. (5) Consequently, departments of local authorities are preoccupied with the delivery of statutory services to new or enlarged deprived groups. The tendency of unemployment to run in families and to be clustered in small local communities (notably peripheral municipal housing estates) adds to the strategic dilemmas that local authorities face. (6) Usually their response is piecemeal and of a 'fire-fighting' nature. Available resources are sparse and, in the first instance, tend to be allocated to highly targeted small-scale projects. Only later is this ad hoc, emergency-type approach replaced by broader, co-ordinated strategic planning and action.

The majority of local authorities in the United Kingdom are at the initial stage of response. This situation arises in part because they lack the necessary structures, resources, and experience to cope. However, they also operate within the constraints of local political and socio-economic cultures. Long-standing acceptance of non-co-operation between departments or with neighbouring local authorities is not swiftly overcome. Also, where local cultures have been built on traditional beliefs about education ending with compulsory schooling and work always being for somebody else, there are few favourable conditions for initiatives based on adult education and training and the principles and practices of enterprise. Thus, at local-authority level, corporate strategies and even targeted areas of strategic planning and action are either promoted or, more likely, hindered by local cultural considerations and experiences as well as by the structures and orientations of their own public services. Clearly the problem is compounded when local government itself is ambivalent about the cultures of adult continuing education and enterprise. (7)

Nevertheless, the development of co-ordinated strategies by local authorities in many parts of the United Kingdom is one of the most unmistakeable trends in the unemployment field. Admittedly few have proceeded further than rudimentary co-ordination, primarily in the economic and educational areas, and full-scale implementation of a corporate strategy has yet to be achieved anywhere. However, sufficient knowledge and experience have been accumulated to provide insights and reference points for those contemplating this kind of development.

This chapter describes and discusses this trend by comparing and contrasting four case studies of co-ordinated strategies evolved by Cheshire County Council, Sheffield City Council, Strathclyde Regional Council, and Wolverhampton Metropolitan Borough Council. These case studies give a broad, if not representative, picture because of variations in size, social cohesiveness, and geographical location as well as in economic, environmental, social, cultural, and, to some extent, political conditions. (8) The common feature of these local authorities is the production of a formal written statement of their strategy. (9) Few other local authorities have reached this stage, but reference will be made to them as appropriate for purposes of clarification and elaboration. (10) The analysis of the strategies will be based on a series of questions. How did the strategy originate? Of what does it consist? How can it be implemented? What are its social and political implications?

## ORIGINS OF CO-ORDINATED STRATEGIES

There is no single shared starting point for developing a co-ordinated strategy on unemployment. The only prerequisite appears to be a situation where long-term unemployment has become a chronic condition. However, within that situation, there are common initiating factors which, whatever their order of priority and appearance, combine to produce that strategy.

Awareness of the gravity of the problem is always prominent. Usually this awareness is couched in terms of 'large-scale and long-term' (Sheffield, Strathclyde) or is designated 'the major social problem of the 1980s' (Cheshire). All four local authorities refer to the relative permanence of the problem ('for the foreseeable future' is the favourite phrase) and note either 'the inability of job creation measures to keep pace with job losses' (Sheffield) or that 'job-creation policies alone are an insufficient response' (Cheshire). Reductions in the area's manufacturing capacity and the unlikelihood of its replacement (as in Sheffield and Wolverhampton) are grounds for this outlook; Cheshire adds the demographic factor of large numbers of young people entering the job market even in the 1990s. Few references are made to the impact on the stock of jobs from the introduction of new technology, although some local authorities hope that this will aid job creation or at least job retention (Wolverhampton).

The severity of the problem is underlined by references to its complexity. Sheffield rejects the notion of 'the unemployed' and argues that there are 'many different groupings'. Clearly unemployment has created new problems or at least exacerbated existing ones in the field of equal opportunities. However, these local authorities do not appear to appreciate fully that the present prolonged period of high levels of unemployment has produced a new social history, and that this new period must be the starting point for strategies rather than former times of full employment. Perhaps the nearest they come to a recognition of this changed situation are references to the inevitable monetary and social costs of having no strategy (Strathclyde, Sheffield). Also missing is an appreciation of the inhibitions imposed by their local cultures, although Strathclyde recognizes its own ambivalence towards the culture of enterprise. (11)

Awareness alone, without factual knowledge and political commitment is, of course, inadequate for devising and implementing strategies. A common source of the latter is a senior political leader. Usually, he is chairman of key committees in the economic and/or educational fields (women rarely fulfil this role, although Sheffield provides an exception in its Employment Programme Committee). It seems that without this political leadership (aided or prompted sometimes by the enterprise of a chief officer), initiatives are either not forthcoming or are not given policy and resource significance.

This leadership depends in turn on information and knowledge that is produced by working parties and/or research units. Wolverhampton is conspicuous in both these respects and has the advantage of a specialized research unit on unemployment at its polytechnic. While other local authorities acknowledge the vital role of research (Cheshire), they have to rely on officer groups (usually wide-ranging in their membership, e.g. Strathclyde) and look to outside research resources or combine forces with them (e.g. Sheffield's Centre of Excellence). (12)

The working parties are characterized in their early stages by sponsorship of a series of unco-ordinated small-scale projects. These projects are targeted at minority groupings and are intended to provide 'first-aid'. Their successes and the clear need to make them more effective by extension and co-ordination stimulate the development of a full-scale strategy (Sheffield, Strathclyde). An example is provided in the chapter by Young. Within the educational field larger-scale projects sponsored by the Department of Education and Science through Replan Education Support Grants from 1985 are performing a similar function. (13) A notable example is Dudley where the project's steering committee includes the chairmen of major committees of the council. At the same time, whatever the source, these projects reveal the inadequacies of existing organizational structures, procedures, and practices (Strathclyde) and lead to plans for extensive reorganization of institutional structures (Wolverhampton).

The outcome of pressure from these various sources is the development of a long-term co-ordinated strategy at local level. What does such a strategy contain?

## FEATURES OF CO-ORDINATED STRATEGIES AT THE LOCAL LEVEL

The linking of 'economic and human dimensions' (Strathclyde) is the basic feature of a co-ordinated strategy. Cheshire refers to a 'blurred relationship' in this respect, whilst Wolverhampton has two separate strategies for these dimensions and only recently has begun to explore their connections. (14) The most fully developed co-ordination appears to be in Sheffield.

Within a co-ordinated strategy the listings of items for action is virtually endless. Strathclyde's 'scope for action' is the most comprehensive and includes poverty, transport, manpower and recruitment policy, purchasing power, health, housing, social work, education, police, leisure and recreation, community development, information and counselling, and unemployed workers' centres. Many of these features are included in other strategies, with health, housing, and police tending to be the most incidental. The most prominent features in all the strategies are information, guidance/counselling, education/training, job creation, recreation, funding, and unemployed workers' groups/centres. Particular attention is given to the following.

**Information** is to be not only of a comprehensive kind but also readily accessible both to unemployed people and to the professionals who work with them. This kind of service is burgeoning, partly aided by the MSC Training Access Points Scheme and usually combining information and guidance services. Sheffield has a well-developed system, and Wolverhampton has planned an ambitious computerized Community Information Service with many local access points.

**Training** The relationship with the MSC is seen as especially significant and problematic. For instance, Wolverhampton notes in its Youth Review that, while the Youth Training Scheme (YTS) 'may mark the beginnings of a necessary and comprehensive training strategy, it loses much of its potential worth because it is not linked to an employment strategy ... without a broadly linked strategy for economic and employment development, the lauded "permanent bridge into work" can strike many young people as a "gang plank to the dole".' However, most MSC schemes are fully utilized by all the local authorities, although there has been considerable opposition to the Jobs Training Scheme (JTS) and the Employment Training Programme because of their 'workfare' implications. Sheffield specifically aimed to make its training for council employees on YTS a model for other employers and also intended to recast the Community Programme to include quality training.

**Job creation** The emphasis in job creation is on affecting the local economy in two forms. One is the sponsorship of self-employment, small firms, and co-operatives (by grants and advice services) and/or development of new products and processes (e.g. Sheffield's Centre For Product Development and Technological Resource). The other is to use the purchasing power of the local authority to the advantage of local firms. Bradford's experience here is illuminating (see the chapter by Kennedy).

**Funding** This element includes MSC, Urban Programmes, EEC schemes, and trust funds; it also covers the funding relationship between departments of the local authority. Sheffield recently examined the latter by means of a general audit of departmental spending on unemployment (it estimates that £8 million out of a budget of £203 million is used for this purpose).

The question of access figures highly in all strategic provisions and raises the issue of costs to unemployed individuals. They are in turn linked to participation in education and training and also to broader mobility questions (Cheshire, Wolverhampton). Strathclyde has used an area-based approach focused on those on low incomes in order to overcome the problem and to increase participation (see the chapter by Young).

Each feature is dealt with in considerable detail in the strategies and often is applied to the needs of distinctive groups. The long-term unemployed grouping is prominent but is added to or sub-divided by women, racial minorities, disabled persons, and youth. The most developed example is that of youth provision in Wolverhampton where the dominant theme is resourcing and enabling young people to control their own situation. Strathclyde and Sheffield have recently stressed consultation with unemployed people over policy matters. (15) However, it is one thing to list and even prioritize constituent parts of a strategy; it is altogether different to co-ordinate them. This problem arises when considering how to implement a co-ordinated strategy.

## IMPLEMENTATION OF CO-ORDINATED STRATEGIES

None of the four strategies for unemployment has been fully implemented. Sheffield is the most advanced, and Wolverhampton and Strathclyde have gone beyond the initial stage. It is clear, however, that gaining approval for and resourcing a co-ordinated strategy are lengthy and highly complicated processes.

The four local authorities identify some reasons for these problems of implementation. First, they make strong reference to the attitudinal change which is necessary. This attitudinal change involves working with the unemployed, not for them (Strathclyde). This approach challenges a 'providers' ethos' and leads to direct resourcing of unemployed workers' groups/centres (Sheffield, Wolverhampton). It also raises questions about the training and consciousness of staff of all grades and kinds. (16) Many dilemmas emerge particularly with respect to changes to traditional professional assumptions and practices within this new context. Individualizing problems in the traditional 'casework' manner and maintaining professional boundaries could be inappropriate and certainly cannot deal with the collective interests and needs which unemployment creates.

Organizational change has to follow. The key area of such change involves improvement of policy-making. This improvement may mean the creation of a policy unit (Wolverhampton) or extending the role of an existing one (Strathclyde). However, it seems certain that committee structures must be reorganized. Sheffield has, for example, formed an Employment Programme Committee which is served by an Unemployment Panel and Unemployment Advisory Group. Wolverhampton has devised a network of working parties (mainly in the educational field) to underpin its strategy, but it too has reorganized its committee structure to include an Economic Development Committee and Committee on Post-16 Education. The process of reorganization and reorientation percolates through to other departments and institutions of the local authority. Colleges and social-services departments appear to be in the forefront in this respect. However, the process is very slow mainly because departments are geared to the efficient delivery of services and not to the management of social change. (17) In general, the managerial issues raised by the strategies (e.g. time-scale or phasing of proposals) appear to be left to officer groups (Sheffield, Cheshire), although Strathclyde advocates an elected members/officers version. This response does not, however, resolve the problem of the form and nature of liaison and co-operation with other agencies, despite a heavy emphasis on collaboration in all the documents. Representation on local-authority committees and working parties appears to be the favourite ploy. By contrast, integrated action with industry seems to be neglected.

Of all the managerial tasks perhaps the least mentioned and practised is that of monitoring and evaluation. There appears to be little expertise and few resources for this activity, and frequently reports are no more than descriptions of developments. Experience indicates, however, that this managerial exercise is fundamental to strategic planning. (18) This weakness is not specific to local authorities, as the chapter by Barber illustrates.

Ultimately the process of implementation depends on funding. The best-laid schemes founder for want of it. It raises questions of knowledge of sources, funding, and expertise in exploiting them, but more perplexing are the political issues of 'internal funding relationship' (Sheffield). Vested interests, of course, do not readily embrace changes that impair their own position. The social and political implications of a co-ordinated strategy on unemployment thus begin to emerge.

## SOCIAL AND POLITICAL IMPLICATIONS

If it is assumed that a co-ordinated strategy can be justified economically, socially, and politically (Sheffield poses the question and affirms that it can be), the most pressing problem for its implementation is whether or not the local authority has the physical capacity and material resources to cope with a successful response. Statutory duties in the field of health care and social services give little scope for additional provisions or reallocation of that provision. Even in the educational sphere, Wolverhampton found that an 8-per-cent response of unemployed people to its Educational Campaign for Unemployed People began to stretch fully its resources of buildings and manpower. Clearly fullest use needs to be made of what other agencies in the voluntary sector and industry can offer. This requirement raises questions about local authorities' relationships with such bodies and organizations, questions that are relatively unexplored in the strategy documents, particularly in the case of industry (including the role of trade unions). (19) For instance, collaboration with industry about community education could alleviate some of the resource problems as well as promote a co-operative spirit, even a nascent communalism.

However, ideological issues are at stake. Sheffield advocates the development of a counter-ideology. It hopes that this response would offset the hold which the present dominant ideology has over many unemployed people and to encourage them into challenging present government policies and actions. This statement betrays the political orientation of Sheffield City Council and suggests that a co-ordinated strategy may be motivated by 'oppositional' politics as well as by the principles of its own political philosophy. Were unemployment strategies deemed necessary when unemployment rose under a Labour government? The question is important because it raises two puzzling issues.

The first concerns the quiescence of the unemployed. Various explanations have been advanced to suggest why possible threats to social order contingent on sharp and continuing increases in unemployment have not materialized. Among them is a reference to the political geography of unemployment which notes its concentration in certain areas that are relatively powerless in the national scene. (20) This explanation needs to be extended to incorporate ameliorating and quietening influences of 'local-authority' responses in those very areas. The irony that these local measures may weaken the will of the unemployed to challenge national policies which prompted the measures in the first place is thereby missed.

The second issue of redefinition of work (including the concept of full employment) and its meanings is more clearly realized by the local authorities. Strathclyde comments: 'A solution is unlikely without a fundamental change in the way society views work and how it should be rewarded.' However, the question of which political agencies could promote this change lies unanswered. (21) The strategy documents are remarkable for their lack of reference to employers' associations and mass media. Essentially they concentrate on reviewing their traditional role. Whilst this review challenges their philosophy and organizational structures and practices, there is little indication how local authorities perceive their role in any major transformation of societal values and structures. Perhaps their political realism over paucity of resources and their competitiveness for what is available limit their political vision. Yet even here there is no recognition that inter-local authority agencies might achieve what they are incapable of doing singly and separately. The political boundaries (and jealousies) of local authorities remain an obstacle to their wider societal role. In any event, the tendency to incorporate and depoliticize local and central government initiatives within the bureaucracy of the state remains the major inhibition to broad social change. (22) 'The ideological battle has been largely won by a view which isolates, separates and stigmatizes people without work ... Is it possible to wage a more successful battle locally? If so, what form should such activity take?' However, even this view appears to be directed more to revising the work ethic than to grappling with the more basic issue of social valuations based on the source of income and the continuing justification of our current wages system (including the trade unions' vested interests in perpetuating it).

However, the immediate major problem for local authorities is whether they can fund their strategies. This issue is the test of their credibility. Ingenuity and determination are plainly required but in the end their persuasive power over central-government policy is critical. The most difficult argument for local authorities concerns securing resources to target provision to combat the uneven distribution of long-term unemployment while high levels of it persist more broadly. Central government's response is an increasing unwillingness to fund individual projects unless they form part of a co-ordinated strategy. This situation may call for novel political alliances and initiatives. Only with such an outcome shall we know if the co-ordinated strategies of local authorities in the United Kingdom can indeed reduce 'the major social problem of our times'. (23)

## REFERENCES

This chapter revises, updates, and elaborates an article by D. Portwood, 'Local authorities unemployment strategies', Journal of Social Policy and Administration, vol. 20, no. 3, autumn 1986, pp. 217-24.

1   In October 1987 long-term unemployed numbered 1,172,000 - 43 per cent of all unemployed people. However, over the previous year, when unemployed numbers fell by 486,000, only 35 per cent of these were long-term unemployed. Reported in The Guardian, 18 November 1987.

2   D. Massey (1983), 'Dimensions of defeat', Marxism

Today, July.

3   S. Becker, S. MacPherson, and F. Falkingham (1987), 'Some local authority responses to poverty', Journal of Local Government Studies, May/June, pp. 35-47.

4   For details of Replan see Uden, T. (1987), 'The Replan experience: a view from the centre', Adult Education, 60 (1).

5   See Association of Metropolitan Authorities (1985), Caring For Unemployed People, London: Bedford Square Press.

6   E.g., Young, R.G. (1987), 'Social strategy in Strathclyde - where now?', Journal of Local Government Studies, May/June, and note the findings of S. Platt on the higher incidence of para-suicide in such communities in Allen, S., Waton, A., Purcel, K., and Wood, S. (1986), The Experience of Unemployment, London: Macmillan, ch. 10.

7   E.g., Young, R.G. (1987), op. cit., p. 14.

8   The political composition of these local authorities is as follows:

| | Cheshire | Sheffield | Strath-clyde | Wolver-hampton |
|---|---|---|---|---|
| Conservative | 27 | 13 | 6 | 24 |
| Labour | 32 | 65 | 87 | 30 |
| Alliance (Lib/SDP) | 11 | 9 | 5 | 6 |
| Independent | 1 | | 2 | |
| SNP | | | 3 | |

The importance of the issue of political diversity is illustrated in an analysis of sets of local economic initiatives in Benington, J. (1986), 'Local economic strategies: paradigms for a planned economy', Journal of the London Economic Unit, vol. 1, no. 1.

9   The strategy statements are: Cheshire County Council (1985) Unemployment: A County Council Strategy, 2 July (extensively reviewed 10 November 1986; specific developments are reported in quarterly newsletter, Prospects in Cheshire, published jointly by the Industrial and Employment Unit and the Cheshire Adult Unemployed Project); Sheffield City Council (1983), Out of Work: Policy Discussion Note, 2 December (this has been updated and elaborated several

times. For a clear statement of their current position see their consultative document, Department of Employment and Economic Development and the Central Policy Unit of Sheffield City Council (1987) Sheffield: Working It Out; Strathclyde Regional Council (1985) Unemployment-Implications for Regional Services (for a commentary on this and on the larger Social Strategy for the Eighties of which it was a part, see Young, R.G. (1987) op.cit.); Wolverhampton Metropolitan Borough Council (1985), Review of the Council's Economic Development Strategy, 22 March and (1985) The Social Condition of Youth in Wolverhampton in 1984 (specific developments are reported in a quarterly bulletin, Economic Review, Policy Unit, Wolverhampton MBC).

10   These include Cambridgeshire County Council, Cleveland County Council, Dudley Metropolitan Borough Council, and Gloucester County Council.

11   Young, R.G. (1987) op. cit., p. 14.

12   A survey of 141 local authorities in the United Kingdom by the Centre for Unemployment Studies, Wolverhampton Polytechnic, revealed that out of 64 replies 35 have research units but these are overwhelmingly focused on economic issues.

13   For details of such developments see Krafchik, M. and Portwood, D. (1987) Developing Educational Provision for Unemployed Adults, Wolverhampton: Centre for Unemployment Studies, The Polytechnic, Wolverhampton.

14   This is common to several local authorities. Benington, J. (1986) op.cit., notes that the bridge in New Left local authorities may be in resourcing trade-union and community employment initiatives (p. 12), but he observes that this is the most controversial area (p. 15).

15   This development is currently in abeyance following a change in political control in Wolverhampton.

16   E.g., Becker, S., MacPherson, S., and Selburn, R., (1983) Saints, Ferrets and Philosophers: Social Workers and Supplementary Benefits, Nottingham: Benefits Research Unit. They noted how social workers are unprepared for welfare-rights work on the scale being demanded. This is a recurring refrain for many professional groups.

17   See Young, R.G. (1987) op. cit., p. 13.

18   See Krafchik, M., and Portwood, D. (1987) op. cit.

19  Sharples, A., (1986) 'The new local economics', <u>Journal of the London Economic Policy Unit</u>, vol. 1, no. 1, discerns a new approach in local government economic policy which desires to cross the public-private divide and intervene directly in the private sector.

20  Massey, D., and Meagan, R. (1983) 'The new geography of jobs', <u>New Society</u>, vol. 63, no. 1061, 17 March.

21  This is the major omission of most of the contributions to 'the future of work' debate, whatever their ideological stance: e.g., Gorz, A. (1985) <u>Paths to Paradise</u>, London: Pluto Press; Therborn, G. (1985) 'West on the dole', <u>Marxism Today</u>, June; Rose, M. (1985) <u>Re-working the Work Ethic</u>, London: Batsford; Watts, A.G. (1983) <u>Education, Unemployment and the Future of Work</u>, Milton Keynes: Open University Press; Handy, C. (1984) <u>The Future of Work</u>, Oxford: Blackwell.

22  Benington, J. (1986) op. cit., p. 23. However, the experience of the Black Country authorities gives an indication of what can be accomplished. See Spencer, K., Taylor, A., Smith, B., Mauson, J., Flynn, N., and Batley, R. (1986) <u>Crisis in the Industrial Heartland</u>, Oxford: Clarendon Press, pp. 158-60.

23  Leaper, R.A.B. (1985), 'Long-term unemployment', <u>Social Policy and Administration</u>, vol. 19, no. 1, Spring.

**Chapter Ten**

## LESSONS FROM STRATHCLYDE'S EXPERIENCE: BOOSTING PEOPLE'S SELF-CONFIDENCE

Ronald Young

> By now it should be understood that Strathclyde Region is not in the business of crash programmes and hopeful rhetoric - but rather of a long-term <u>process of transforming the way people think about themselves and what they are capable of and of reshaping our methods of implementation accordingly</u>. (1)

In 1976 Strathclyde Regional Council identified multiple deprivation as its major problem. For a major local authority to admit that there was a problem of urban poverty and ghettos was then a major achievement in itself. Although a lot of people in local government had been denying that such problems existed, several national reports in the early 1970s demonstrated vividly the extent to which youngsters in run-down urban areas of the West of Scotland were 'disadvantaged'. (2) Indeed the establishment of Strathclyde Region in 1974/5 was seen as an opportunity to make the break with a local-government machine that was seen by some in the new council as insensitive, paternalistic, unequal, complacent, and repressive.

Unemployment was both high and long-term in many council-house estates and, therefore, new jobs formed a very relevant theme. But simply to call for more jobs was little more than a fatalistic restatement of the problem. It was, after all, precisely these areas that were left high and dry when the national economy reflated. Something had to be done directly to remove the stigma and the feeling of cynicism and hopelessness so evident in these council-house estates. For those behind the Strathclyde Region's new policy, there was also concern about the way the local-government machine operated (deliberately or otherwise) to repress the hopes of ordinary people. The problem was essentially how people on low incomes (in both the inner city and the outer estates) were treated and made to see themselves and their capabilities - 'living down' to other people's expectations.

The strategy involved 'changing the way (disadvantaged) people think of themselves and what they are capable of'. (3) It meant operating at two levels: boosting the self-confidence of those whose expectations had been lowered - and in so doing challenging the myths which abounded about the competence of so many ordinary people; (4) and changing those aspects of the operation of public agencies that led to discrimination and poor services in the field of health and housing for low-income people.

The last twelve years can, with the benefit of hindsight, be divided into three phases. The first phase - from the initial policy statement until the early 1980s - involved introducing three new elements into Scottish local government: positive discrimination - trying to ensure that services in 45 designated Areas for Priority Treatment (APTs) were at least as good as elsewhere; community action - resourcing local groups to take initiatives; and advocacy - using the clout of the Strathclyde Region to change the policies of health, housing, and welfare agencies. In 1981/2 the strategy was intensively reviewed (by a range of people, not least community activists themselves, at six major community conferences), and the result embodied in the distinctive document Social Strategy for the Eighties. This strategy document concluded that, although an area-based approach was still needed, this approach should be balanced by developing such programmes as 'pre-fives'; community business; adult education; services to the unemployed; and stronger area initiatives.

The second phase (between 1981 and 1986) was concerned with the implementation of the Social Strategy for the Eighties. The period 1986/7 saw another process of stocktaking, heralding a third phase of what is a unique local effort to tackle the problems of the long-term unemployed. This effort attempts to integrate social and economic strategies at a local level.

The aims of this chapter are to outline the work of the past decade, and to try to indicate what lessons can be drawn from the Strathclyde experience, particularly in relation to the dynamics of individual and organizational

195

change. In particular it asks:

What has the Strathclyde Regional Council been trying
to do?
What precisely has it been trying to change?
What methods has it employed?
When are results expected?
How are they to be measured?
Why does Strathclyde Regional Council keep re-
inventing the wheel?

The background to this strategy is provided by the sheer
scale of the Strathclyde Regional Council, making it
distinctive within the overall UK context. It employs
103,000 staff and has a budget of £1,700 million for a
population of 2.5 million. Its responsibilities cover
education, social work, roads, police, water and sewerage,
economic development, consumer protection, and strategic
planning (but not housing or health).

## WHAT HAS THE STRATHCLYDE REGION BEEN TRYING TO DO?

The 1976 document on urban deprivation was written (after
some twelve months of anguished discussions) to spell out
the sort of action that could be taken at a local level to deal
with urban poverty. It was also challenging various myths:
e.g. that Strathclyde had to wait for central government to
improve the national economy; that the long-term
unemployed had only themselves to blame; and that all that
was essentially needed was £500 million (say) of capital
allocation to be given to local authorities for house building
and other environmental improvements. The 1976 document
on urban deprivation and the work that flowed from it was
trying to do four things:

put a new issue on the agenda: namely, the way that the
labour and housing markets and public services
stigmatized and discriminated against those who lived
in certain municipal housing estates. There were strong
vested interests trying to deny this situation.

assert the rights and capabilities of those people who
had been labelled as second class. Strathclyde Regional

Council was concerned about the way that the local-government machine operated (deliberately or otherwise) to repress the hopes of ordinary people.

encourage people in those areas to organize and take action. There was a feeling of hopelessness which experience had demonstrated could be broken.

change those aspects of the operation of public agencies which led to discrimination and poor services in the field of health and housing for low-income people - and generally make public agencies more responsive.

## WHAT METHODS HAS STRATHCLYDE REGION EMPLOYED?

This strategy represented an unusual venture for a UK local authority. Most unusual of all, however, were the methods employed to give effect to the strategy. The conventional model of decision-making is for politicians to pass specific laws which are then implemented by the managers and professionals. Such a system of decision-making flows from the belief of the politicians in their own and their advisers' wisdom, competence, and authority. In 1976 Strathclyde Regional Council questioned these beliefs, certainly in relation to social deprivation and polarization. Neither the Regional Council nor its advisers had much clue about the nature and causes of, and solutions to, multiple deprivation. It could, however, see the damage that the autocratic or 'top-down' model of decision making had caused - amongst both staff (cynicism) and clients (dependence). To counteract this damage, four approaches have, at different times over the past 13 years, been adopted: resourcing local action; putting the local-authority house in order; advocacy; and developing certain neglected programmes.

### Resourcing local action

Strathclyde Regional Council's first approach has been a community development, namely encouraging local people to take initiatives by:

appointing community work staff to the designated

APTs (which increased in number from 45 in 1976 to 88 in 1981). The staff were essentially in the Social Work Department, although certain community-education staff also had relevant skills. More than 250 such staff were in place by 1988.

making resources available to local groups, essentially through the Urban Programme, which funds 2,000 staff and some 600 projects. Some £4-5 million a year has been available for new starts.

setting up in 1978 a major Community Development Committee to ensure that issues of concern to local communities were taken seriously. From the start, the style of this committee was deliberately different from that of the traditional committee, with many member/officer groups being established on issues such as youth work, ethnic minorities, and play-schemes.

encouraging local structures in these APTs. Each of the 88 designated APTs was to have local strategy groups. These groups are essentially forums chaired by a regional councillor and serviced by a regional community worker with the broad objective of improving local services, whether by bids, advocacy, or self-help.

Community development on its own was, however, recognized as insufficient; indeed, it could be counter-productive if community initiatives hit an indifferent or hostile local government machine. Hence a second approach was adopted in parallel.

## 'Putting our house in order'

This approach meant trying to ensure that resources were allocated on the basis of need and that services were made more accessible and responsive. Joint area initiatives were undertaken in conjunction with six district councils in seven neighbourhoods between 1978 and 1983. They were headed by specially appointed co-ordinators with the aim of developing appropriate local initiatives and, in so doing experimenting with more decentralized forms of decision making. In the light of this experience - and that from 198 with the joint economic initiatives involving both the Scottish Development Agency and the districts - the Council has (from 1984) negotiated six Special Joint Social and Economic Initiatives in the larger housing estates that have both economic and social objectives.

## Advocacy

Despite the Strathclyde Regional Council's size and budget, it has no responsibility for housing, health, or welfare payments - crucial elements in the cycle of urban poverty and deprivation. Quite explicitly, from 1976, the Regional Council had, therefore, been in the business of attempting to change the policies and practices of other agencies: for instance, on dumping policy, on monolithic municipal tenure, on paternalistic health practices, on stigmatization of those on welfare, and on low take-up of welfare payments.

Strathclyde Region led the way in the 1970s on welfare benefit campaigns and by 1988 employed 50 welfare-rights advisers, plus many more information and advice workers. Approaches to health and housing have been more subtle and therefore long term. Unfortunately, government changes and hostility have knocked more sense into housing management than local pressures and initiatives (except for Glasgow).

## Programme development (since 1981)

The whole thrust of the Social Strategy for the Eighties document was to balance the area concern of the 1976 strategy with a programme focus. There has always been a dual meaning to 'community' - not just a geographical neighbourhood but a 'community of interest'. On the one hand, the stigmatization and neglect of housing schemes in the late 1960s and 1970s justified the 'area' emphasis of the 1976 strategy; on the other, the 'neighbourhood' is too amorphous and diverse a concept to constitute the exclusive basis for a change strategy - as is evident in the strength of the Sleeping Giant forums developed by senior citizens particularly in the Glasgow area in the mid-1980s, and the amazing growth in that same period of 'under-fives' link-up groups (by 1988 some 100). However, both the 'areas' and the programmes' required more effective organization.

This perception lay at the basis of the recommendation of a 'local social-strategy group' for every APT (chaired by the regional councillor). The recommendation drew on the experience of the 'under-fives' member/officer group which had used local information on under-fives' needs to select twelve areas and to invite local staff and parents to come together and bid for extra resources. The relationship between the Divisional Deprivation Group of five to six councillors, established in 1981, and the local groups was seen as crucial in the development of a similar process of negotiating change in the key policy areas. (5)

The key components of programme development were:

the 'pre-five' service (the establishment in 1985 of an integrated 'Pre-Five' Committee and Unit and the spontaneous development of more than 100 local link groups of parents and staff).

adult education (a variety of autonomous local programmes).

community business (the establishment in 1982 of an independent agency to encourage local employment initiatives in APTs).

information and advice (establishment by local people of local centres).

the Unemployment Initiatives Unit (established in 1987 and stemming from the recognition in 1982 of the new world of high and long-term unemployment).

The Unemployment Initiative Unit, with eight staff, developed work in three broad areas - all underpinned by the related themes of co-ordination and corporate working. These areas were:

comprehensive information services to those unemployed: e.g. the 1-in-4 Newspaper delivered to every household, booklets, leaflets, and an innovative viewdata service, now being piloted. The latter development will, when fully deployed, involve the siting of over 100 viewdata units in a wide range of local outlets in Areas of Priority Treatment in Strathclyde.

training aimed at Council and other agency staff: examples of the new issue- and project-development-based training include unemployment, health, and local social and economic project development.

policy development by the Unit and a newly created Unemployment Issues Steering Group, membership of which included elected members, senior officers of the Council and other relevant agencies (i.e. the voluntary sector, the Federation of Unemployed Workers' Centres, and district councils). The Steering Group examined ways of taking forward recommendations on manpower and recruitment, purchasing, service delivery and adaptation, to practical programmes for action, using a cross-committee/department/agency approach.

## WHAT RESOURCES HAVE BEEN DEPLOYED?

'Bonding' mainline programmes to the new strategy was not easy, particularly when Regional Support Grant was being dramatically cut. The crucial resource for the strategy was in fact £4-5 million of new annual development grant from the Urban Programme. In relation to the Strathclyde Region's £1,600 million budget this amount was obviously miniscule, but it was virtually the only development money. The Urban Programme had several distinctive features: it was responsive: local people design the projects; it was innovative: new ways of designing and delivering policies were being encouraged; it was temporary: which meant that at the end of the funding period it is subject to far more rigorous appraisal than normal local-authority activities and services.

Unfortunately, central government introduced two new features in the 1980s. The programme was made competitive: it was cut during the 1980s from the £40 million promised in the 1982 White Paper on Public Spending to a £30-million level in the mid-1980s (it should be at least £50 million if the 1977 White Paper on Inner Cities were being implemented). In the summer of 1986 central government indicated that it 'anticipated' a switch from those authorities which had been intensive users of the Urban Programme to other local authorities which were increasingly bidding for the money. Although the Scottish Office approved the projects, government denied that this was a policy - simply a trend! However, during 1986/7 Strathclyde was allowed to spend only 30 per cent of its allocation. It was not clear where the missing money had gone. Also, the programme was made punitive: as central government funding comes to an end and local authorities pick up the pieces, the latter 'overspend' and incur clawback penalties. In Strathclyde's case this amounted to more than 100 per cent of the value of the projects.

**Table 10.1** Types of Urban Programme projects in Strathclyde Regional Council

| Client group | No of projects |
| --- | --- |
| Under fives | 82 |
| Young people | 78 |
| Elderly | 23 |
| Single parent | 8 |
| Offenders | 10 |
| Alcoholics | 23 |
| Physically handicapped | 12 |
| Mentally handicapped | 20 |
| Education | 25 |
| Employment | 42 |
| Community buildings | 36 |
| Community services | 219 |

As Strathclyde Regional Council wrestled with the host of difficulties generated by the Urban Programme, there were times when it was tempted to abandon its interest. It was difficult to take seriously a programme whose (small-scale) projects:

had an average of 15 months between conception and start.

had the most ambitious of aims being tackled by the most junior staff.

who were then left unsupported by the apparatus of local government.

ran for four years but in the final year were subjected to stricter scrutiny than any other part of the local authority (not least because the cost to the local authority of 'picking up the tab' was a ten-fold increase (because of penalties) in its expenditure).

Even so, the Urban Programme was worthwhile because it generated - by the grass-roots origin of the innovative projects it encouraged - a far more participative style of programme than that for which local government in the past had shown enthusiasm. The financial carrot of the Urban Programme was needed to bring forward projects for and from certain neglected client groups and areas, to be managed in a different way.

Some argue that the Urban Programme should be abolished and urban local authorities given a proper Rate Support Grant weighting and level of funding to allow an adequate provision of services. This argument betrays a naive confidence in the capacity of local bureaucracies to deliver the goods. Certainly Strathclyde's whole social strategy assumption and experience demonstrates the dynamic effect of a design process in which the local authority does not have a monopoly - but in which the community, the local authority, and an outside agency have to negotiate.

## WHAT PRECISELY HAS STRATHCLYDE REGION BEEN TRYING TO ACHIEVE?

Strathclyde Regional Council's strategy focused on the removal of certain negative features in local-government processes and on the achievement of certain goals. Two negative features were attacked. First, the strategy challenged the way in which the hierarchical system of bureaucracy undermined local initiative, both of staff and of the community. As the Social Strategy document put it, 'The majority of staff are discouraged from joint work with councillors, other professionals and residents in these areas by the way the traditional department system of local government works. Career advance depends on one's work as a professional or manager in a particular department and not on the collaborative ventures emphasized in this document and the earlier one of 1976. That is the crucial issue that must now be faced and resolved. Exhortation and good intentions are no longer enough'. (6) Second, an effort was made to change the judgements held by many staff about the capabilities of the residents of APTs. The positive goals were to improve accessibility to services; increase opportunities to develop skills and access job opportunities; improve self-image by self-help activities; and increase resources (both collective and individual).

## WHEN WERE RESULTS EXPECTED?

'One-shot, one-time programmes will have limited effects, especially with the poorest of the poor. Social policy should not be orientated to dealing with what are considered residual problems in a short time span'. (7) Tackling discrimination and injustice is, as Strathclyde Regional Council always recognized, a long-term process - particularly when the means being used are themselves so much part of the problem; resources are being cut back; only a small minority of people are engaged on the venture, whether at the centre or at the periphery; and the programmes are experimental. Strathclyde spoke informally about a 20-year process.

At the same time, if Strathclyde were to generate a momentum and a feeling of confidence, some quick 'successes' were needed. Hence the Regional Council looked to its Area Initiative to choose one or two 'manageable' problems in the first instance.

## HOW WERE RESULTS TO BE MEASURED?

No measures of results were specified in 1976; after all, Strathclyde Regional Council was wrestling with a very new venture. In the 1982 Review (recalling that Strathclyde was attempting to boost self-confidence) the Regional Council suggested measures of results in two areas:

**Community action** How many more people are now prepared to fight for their rights? Are they more knowledgeable; are they being helped by the local authorities? Do they have access to appropriate advice and resources?

**Take-up of services** Are resources going to those who most need them? Simply looking at the overall numbers taking advantage of apparently relevant services such as adult education can be misleading.

As the numbers out of work continued to increase, Strathclyde expected those taking advantage of local-authority services to increase as well. More worrying, those taking advantage of services might be (and generally are) those already with the relevant skills and confidence. To get round these difficulties required not merely the needs/resources analysis already carried out by the Social Work Department but social-survey work of a sort not so far undertaken by the Region - to identify more clearly the target groups, the difficulties that they face, and their experiences as they get involved in local-authority services. (8)

These two measures of results are somewhat general and liable to be dismissed by the technocratic wizards of performance review who are more interested in hard, 'objective' data. However, this strategy has been about the boosting of people's confidence: the two measures suggested in 1982 are directly relevant to that.

The trebling of unemployment during the 1980s has brought this particular blight home to the vast majority; in that sense, notwithstanding the efforts of central government to the contrary, the image of 'the scrounger' no longer serves to stigmatize the long-term unemployed. This shift has undoubtedly made the basic aims of the strategy - giving dignity and skills as well as services to the stigmatized - less difficult. But it has brought with it new problems. Certain groups are particularly vulnerable during such unemployment, women and youngsters in particular. In this respect the nature as well as the scale of the problem has been changing. This development makes it all the more important to have precision of targets. In their absence cynicism and hopelessness will prevail.

Other measures that Strathclyde might have suggested are:

**Continued resource input** Here two quotes are relevant:

> While the complaint is often made that the poor are handicapped by a presumably short time span, they are

> frequently more handicapped by the short time span of public policies as policy attention wanders from one issue to another.
>
> A programme is what it does, not what it would like to do or was established to do. The distribution of funds and staff time are good indicators of what an organization actually does rather than what it believes it does or tries to convince others that it does. (9)

All too often policy 'wanders': a policy has no sooner been set up and announced than the policy-makers are assailed by another crisis or by the routines of their departments. A new flavour of the month emerges. Quite genuinely, policy-makers imagine that the first policy has now been implemented, not only with the intended financial resources but, often more importantly, with the same zeal with which they themselves designed the policy. In most cases they could not be more wrong! That essentially is why structures are needed - as permanent reminders of historical commitments. But more is needed: in particular, measures of the amounts of energy, time, and money that are going into a given policy area (compared with the others).

**Rate of improvement in the designated areas,** measured by indices which can be controlled at local level, such as vacancy rates, educational achievement, and popularity of scheme. How much is known about these processes of urban decline and regeneration? How systematic is that knowledge? And how much is it used in policy decisions? When the Region was selecting in the early 1980s from its 88 APTs the Special Joint Initiatives, it did so partly on the basis of an assessment of the social dynamics. However, if we are in the business of social engineering, it is necessary to become somewhat more sophisticated in assessment of group dynamics and more ready to adjust interventions accordingly. At another level, there has been a national consensus that Glasgow has dramatically 'picked itself up by its bootstraps' and laid claim to being 'Britain's first major post-industrial success'. (10) But do we really understand that process in a way that can sustain the development? In Generating Change, a document the Region issued in September 1987, it tried to spell out some of the lessons of a decade of successful urban regeneration work. Even so i was still a very tentative effort.

## SOME DILEMMAS OF SOCIAL REFORM

The dilemmas posed by Strathclyde Regional Council's approach were numerous: e.g.

### 'Area' versus 'client' focus

Apart from the argument about 'need' existing outside APTs (and the danger that success would merely make APTs of areas currently at risk), how did Strathclyde know that the resources were going, within the APTs, to those most in need? And was that, in any event, its aim; or was it to boost the morale of the average upwardly mobile person in those areas? The tension between 'area' and 'client' was reflected in organizational strain, with officers increasingly subjected to pulls from their departmental structure <u>and</u> neighbourhoods.

### The 'bottom-up' versus 'top-down' approach

Some impatience developed in the early 1980s with the pure community-development model. Rather cruelly perhaps, the approach was once designated as 'drawing lines round certain areas, saying we could do better, dangling an Urban Aid carrot and waiting for something to happen'. (11) It is certainly at odds with canons of good management which want goals and programmes. Even so, there has been method in the madness. The exploratory nature of the Strathclyde strategy has made it difficult to dismiss it as a political dogma imposed from on high. The strategy has reflected basic principles of common justice and has been flexible enough to leave a lot of scope and allow a sense of ownership of the strategy to develop. The <u>Social Strategy for the Eighties</u> document has also given the necessary support to reformist officers when they encountered bureaucratic inertia or professional suspicion about the unorthodox ventures into which their commitment to community development got them.

A new model has emerged in the field of 'under fives'. Here a detailed and authoritative statement (on the lines of the Region's 1976 document on deprivation) was the subject of a sustained dialogue between senior councillors and officials, on the one hand, and local practitioners in twelve areas that were selected as being particularly deficient, on the other. Staff and parents in those areas were invited to indicate how the situation could be improved, whether by new practice or by extra resources. This process has demonstrated appropriate roles for headquarters and localities within the confines of a single-issue approach.

### Single-issue versus the all-embracing approach

It might seem that attacking on a single issue - whether incomes, house modernization, community facilities, nursery provision, etc. - is more likely to generate confidence (and results) than a complex, all-embracing strategy. At the same time, the services and actors interact and cannot be divorced from one another. As S.M. Miller put it in a keynote review: 'People live in communities, in groups, in families. Programmes cannot successfully help them if they are treated as atomistic individuals'. (12)

### Corporate consensus versus advocacy

To what extent do people seriously expect real change to come from working groups that are made up of different agencies or departments? Organizational as well as political history tells us that it is more likely to come from conflict - in the sense of a small group of people trying to impress their particular perception or argument on others.

### Collective versus individual motivation

The support of a group is crucial to the development of demoralized people: often, however, the stage is reached when the group can be constraining rather than liberating.

### Realism versus hype

How a programme starts is important: what it promises, the expectations that it raises. The poor are frequently both suspicious and deceivable - expectations can rise very rapidly and collapse suddenly. (13)

Initially a low-key approach was adopted in Strathclyde: no trumpets were blown, no promises made. The only appropriate one would indeed have been Churchillian. In fact, because the aim of the initiatives is boosting self-confidence - individual or collective, it requires publicity and publicity material of a particularly sensitive sort, controlled very much by local people.

## Need versus opportunity

If one is in the 'confidence' game (boosting it), one does not initially take on the most difficult problems. Nothing, after all, succeeds like success. Generating and pursuing development are difficult processes: they involve a tricky balance between social and commercial purposes.

## THE ACHIEVEMENTS OF STRATHCLYDE REGIONAL COUNCIL

The Strathclyde Region can produce a lot of impressive statistics and facts to show reasonable endeavour in the face of increasing difficulties, both for itself as an organization and for an increasing number of unemployed. This evidence should, however, be tested against the four measures suggested on pages 204-6.

### 1    Community action

There has been very encouraging work in the fields of credit unions, play-schemes, community business and health projects. However, here the results might have been even more encouraging. Some truth remains in the observation in the Social Strategy for the Eighties document that:

Even then we had examples of good practice developing in some localities by dint of individual officers or politicians being prepared to try something different. In some places it had been police initiatives; in others adult education, and in yet others health initiatives. In many cases, these were accompanied by rapid movements in the indices of social malaise. The tragedy, however, is how isolated such simple initiatives were - and how many obstacles seemed to be placed in their way.

## 2    Take-up of services

The most dramatic changes have been in adult education, with 6,000 adults returning to day-time classes in schools and a similar number using the Open University courses. The same trends in service use can be seen wherever local offices have been opened, e.g. the Ferguslie Park Initiative.

## 3    Resource input

Resource input has generally remained static, if not declined, in relation to increased demand and expectations (e.g. in the 'pre-five' field). Resources have, however, begun to flow to the larger of the Special Joint Initiatives.

## 4    Neighbourhood revival

Some neighbourhoods (like Maryhill, Barrowfield, and Northwest Kilmarnock) seem to have acquired a positive reputation by virtue of a combination of social, employment, and environmental development processes. So far, however, there does not appear to be a winning formula - let alone one that agencies are capable of applying. Clearly factors such as luck and determination come into play.

By 1988 - in the absence of comprehensive survey work - Strathclyde Region had no real answers on measures 1, 2, and 4. The Region did, however, pilot a social survey in a couple of areas.

Some questions have to be asked about:

the balance of time spent by senior people on social development issues (compared, for example, with the defensive work in relation to central government).

the extent to which the traditional operations of local

government, so criticized in the 1976 document, have actually changed (see above).

whether major upheavals - such as the restructuring of the Social Work Department or the major exercise on secondary and primary schools' rationalization - have helped or hindered the strategy.

## WHAT MORE SHOULD BE DONE BY LOCAL AUTHORITIES?

Local authorities must do more to set targets; to resource strategy (with more than funny money); to ensure that other policies and services are appropriately adjusted; to monitor implementation; to give proper support to change agents and community leaders; and to understand and encourage community enterprise.

### Detailed targets can and must be set

'Ambiguous, conflicting programme goals and activities lead to trouble. Most programmes have this problem'. (14) In the 1982 Social Strategy document Strathclyde Regional Council recognized that 'many staff did not know we had a strategy, let alone what it meant for them' and anticipated the production of detailed service objectives and targets that would translate the rhetoric into management tasks. This process never happened. The task is, arguably, beyond the capacity of any headquarters. It is something that must be negotiated on the basis of each area office's initial definition. It is hoped that the new District Social Work Plan process will have a relevance here; also the local review structure set up as part of the schools' review should have this sort of focus. However, such decentralized processes require two further elements: an overarching philosophy which can only come from a combination of leadership and 'internal' marketing/training of a sort pioneered recently by British Airways; and more of a team approach at a local level. A challenge to departmental boundaries is needed and is missing, apart from the impact of local strategy groups.

*Combating Long-term Unemployment*

## Programmes must be properly resourced

> Organisation is fateful. How programmes are organised
> affects what happens to those who deal with them.
> Where programmes are aimed at the short-run, are
> characterised by uncertain funding, high staff turnover,
> and poor planning and organisation, it will be difficult
> for people to accept or benefit from them. (15)

The Urban Programme has tended to absorb the time of
senior people to the exclusion of their consideration of other
elements of the Social Strategy, such as mainline spending
and policies. The Divisional Deprivation Groups (in 1981) and
the Divisional Community Development Committees (from
1986) were established to ensure a focus on the mainline
departments but were not resourced to produce alternative
approaches, let alone implement them.

There is something very dubious about the allocation of
central tasks to peripheral structures and individuals. As
long as the councillors involved in such area-based work are
junior councillors, and the relevant political career is seen
to be that of heading a committee, then one cannot expect
area-based work to achieve major change within the local
authority.

## The management system should reflect the strategy

The various initiatives have been grafted on to the existing
committee and departmental structures and cultures. The
basic problem is that local-authority departments - like
leopards - seem unable to change their spots. Committees
remain the creatures of their department: the brave hopes
of the 1970s for policy committees, chief executive
departments, and even performance review have been
frustrated. Strathclyde's Social Strategy document warned
of this danger in the following quote from Robin Hambleton:

> Approaches like the American Model Cities Programme
> and the current British Inner City Initiative, by
> attempting to build positive discrimination in favour of
> specific areas into existing services, by insisting on a
> more co-ordinated approach to the problems of these
> areas, and by trying to open up the processes of
> decision-making, challenge fundamental organizing

212

principles of urban government - uniformity of service provision, functional service management, and formal political and departmental hierarchies of control.

In these circumstances it is inevitable that new initiatives will be faced with formidable opposition from entrenched interests. Whilst some opposition may take the form of hostile resistance, a more subtle and probably widespread response is to absorb the threat - to defuse, dilute and redirect the energies originally directed towards change. (16)

## The strategy should be properly monitored

Information is power. It is only in the last few years that information has been collected systematically about how the local-authority resources in areas of priority treatment relate to the needs that have been identified. Without such information any strategy is just pious good intentions.

## Support change agents

It is a truism to say that we are in a time of rapid change: a lot of lip service is paid to the need for retraining. Very often, however, it is precisely those in the crucial positions of managing or obstructing that change who lack appropriate opportunities for training. At least three levels can be identified: political, managerial, and community. No self-respecting private company would introduce new products or systems without massive training. More progressive companies will pull in business schools and even set up, with their support, a teaching company. The time is overdue for such an approach in local government.

## Understand and encourage community enterprise

Since the late 1970s Strathclyde has seen a lot of inspiring work by unemployed people in its APTs. Also there have been encouraging developments like resource centres to ensure that such people get appropriate technical assistance. However, there is a long way to go before the unemployed get proper access to existing local-authority services, let alone have available appropriate services.

213

'Community enterprises' are a diverse group in terms of their objectives and their activities: some produce goods; some provide services to meet local needs; some are concerned with training people in new skills; some provide workspaces to enable people to go into business; and some promote creative activities without a direct focus on providing employment.

Some would argue that it is naive to expect such effort to produce much change; they would echo Tolstoy in suggesting that the strength of attitudes and of vested interests of bureaucratized professionalism is so great that only marginal and temporary concessions will ever be made. Others would go even further and argue that real self-confidence will only be achieved when individuals are autonomous, i.e. the market solution. There is certainly more in this argument than most Left-wing people would concede: equally, the Right has to accept that the narrow Thatcherite definition of the market ignores the potential of the non-profit-making 'third' or community-enterprise sector.

The realization that the high levels of unemployment in some local authorities will not be overcome either by traditional methods of intervention or by commercial enterprise necessitates a more radical rethinking of our social and economic policies, particularly in relation to the concept of work. There is still far too much confusion about what constitutes 'real' jobs: some people think that term covers only manufacturing industry, others anything that makes a profit. Behind this confusion lies many an ideological posture.

Behind the diversity of their activities most community enterprises have certain key features in common: they focus on promoting new enterprise activity by people who have limited previous experience; and they aim to build up self-confidence and skills to enable people to take initiatives on their own behalf. The cost effectiveness of work by community workers and lay activists is immense; the financial return to a community from such work is equally large. However, these factors do not seem to be properly appreciated. Sometimes the worth and significance of the work undertaken in these neighbourhood projects are not even properly appreciated by those doing it. They have to battle so much not only against indifference but indeed against obstacles put in their way by the rules and regulations of various agencies. Consequently they tend to absorb the negative perceptions of officials who get rather fed up with dealing with approaches that cannot fit neatly into the bureaucratic rules.

At the moment the only support for community enterprises comes from an ad hoc collection of community workers and volunteers who may have had training in group skills but have little background in such things as accounting, management, and marketing. In the 1980s the fashion for small business has spawned a network of support mechanisms; it is time to recognize that concepts such as community enterprise and community care demand the same provision of innovative, managerial, financial, and technical skills and support.

The evaluation of Strathclyde Community Business published in March 1988 started a debate about the way forward, building on the achievements of the past decade. Already a consensus seems to be developing that there should be a more 'federal' structure of local community-enterprise development agencies - the 'community' equivalent of local enterprise trusts.

## CONCLUSIONS BASED ON STRATHCLYDE'S EXPERIENCE

Nothing seems to change in local government. Initiatives are still made to jump through hoops; groups go on reinventing the wheel. Change occurs not because of local-government processes, policies, and cultures but despite them. It owes much to obstinate, idealistic individuals who refuse to take no for an answer. Why should this be so? And can it be changed?

The constraints on action at the local level are powerful and include:

### The 'segmentalist' culture (17)

This phrase is used instead of the more obvious and emotive ones of 'departmental', 'bureaucratic', and 'professional' to describe (rather than vilify) an organizational construct which has been questioned for the past two decades. However, despite all the brave rhetoric about management revolutions and decentralization, a 'segmentalist' culture has continued to exist in local government. (18)

215

## The political values embodied in British local democracy

Parochialism produces a failure to appreciate that, in European terms, Britain is somewhat unusual in certain aspects of its local politics and government. Two features are especially notable. One is that the diffusion of local-authority responsibilities is arguably so wide as to reduce politicians to a reactive managerial role. In many other countries local politicians operate as the catalysts of change locally in a pluralistic rather than monolithic institutional setting. (19) The other is that lack of leadership is an institutional and not just a personal factor. In the British tradition the line between 'administration' and 'politics' is sharply drawn and embodied in the dual leadership roles of leader of the council and chief executive. This situation contrasts with, for example, the German pattern. (20)

## The attack on local government from central government in the 1980s

This attack has not only distracted the time and energies of key people in local government but also made them defensive and hesitant about giving hostages to fortune by being critical of organizational performance themselves.

New possibilities can only be opened by a determined effort to bring the feudal empires that are local-authority departments under proper control. Clear targets, priorities, and annual reports from the directors on performance are a necessary start. At stake is not just whether local authorities can do more but whether local government wants to survive. Survival will involve a redefinition of the role of local government and of its councillors and officials. British local government is moving towards a more enabling role - and that is no bad thing. The corollaries are that the role of councillors should become less managerial and more concerned with the dynamics of change and influence; and that officers are placed in more action-orientated than bureaucratic roles, probably in local development-agency type work and matrix rather than departmental structures. There is a need for fresh concepts and perspectives that take as their starting point not current vested interests or the tired clichés of party ideologies but the need to empower significant numbers of disadvantaged groups in a rapidly changing society.

# REFERENCES

1 Strathclyde Regional Council (1982) <u>Social Strategy for the Eighties</u>.
2 Principally National Children's Bureau (1973) <u>Born to Fail?</u>, and Department of the Environment (1975/6) <u>Census Indicators of Urban Deprivation</u>.
3 From para 11.2, page 49, of <u>Social Strategy for the Eighties</u>.
4 One early example of such demystification was Sheila McKay (1980) <u>View from the Hill</u>, Local Government Unit.
5 From R. Young (1987) 'Social strategy in Strathclyde - where now?', <u>Local Government Studies</u>, May/June.
6 From <u>Social Strategy for the Eighties</u>.
7 S.M. Miller (1980) 'Reinventing the broken wheel', <u>Social Policy</u>, vol. 10, no. 5, pp. 2-3.
8 For some examples see National Consumer Council (1986) <u>Measuring Up</u>.
9 For a critique see I. Hoos (1972) <u>Systems Analysis in Public Policy: A Critique</u>, Berkeley: University of California Press.
10 From special article in <u>Sunday Times</u> 2 December 1984.
11 In 'Decentralization in Strathclyde', <u>Local Government Policy-Making</u>, July 1984, p.27.
12 S.M. Miller, <u>op.cit.</u>
13 ibid.
14 ibid.
15 ibid.
16 R. Hambleton (1981) <u>Policy and Politics</u>, vol. 9.
17 The phrase is Rosemary Kanter's in her masterly and neglected book (1984) <u>The Change Masters</u>, London: Counterpoint.
18 For the latest statement see John Stewart (1986) <u>The New Management of Local Government</u>, London: Allen and Unwin.
19 See (1986) <u>The Widdicombe Report</u> London: HMSO, and the arguments in 'Community development - the political administrative challenge' in P. Henderson and D. Thomas (eds.) (1981) <u>Readings in Community Work</u>, London: Allen and Unwin.
20 Not for nothing was local government called by a sympathetic observer, 'The headless state - the non-accountable executive' (Professor D. Regan).

# Chapter Eleven

## LESSONS FROM BRADFORD'S EXPERIENCE: THE INTEGRATED APPROACH

David Kennedy

This chapter outlines the context and nature of unemployment in Bradford and then analyses the origins and character of Bradford Metropolitan District Council's strategy to combat unemployment and restructuring of its services in order to implement this strategy. Bradford is interesting as a case study of an integrated approach to combating unemployment in which the Council sees a key role for the European Commission.

## UNEMPLOYMENT IN BRADFORD: TRENDS AND INCIDENCE

Unemployment is a major component of deprivation and disadvantage for the people of Bradford. The January 1988 unemployment figures showed nearly 25,000 people out of work: one in eight of the Metropolitan District's workforce. If we were to include people who are looking for work but not eligible for benefit, this figure would increase considerably. This chapter looks at unemployment from the standpoint that the unemployment figures in themselves are not the main problem; the real issue behind unemployment for Bradford's people is the poverty and hardship that unemployment brings (see the chapter by Benington). This distinction is not just an academic one - it has very real implications for the Council's analysis of the District's unemployment and for the nature of its policy responses.

The January 1988 unemployment statistics showed that 24,737 people (17,872 men and 6,865 women) were unemployed in Bradford - an unemployment rate of around 12 per cent. From a relatively low 6.5 per cent in October 1979 the unemployment rate for Bradford reached 14.8 per cent in October 1981 as massive redundancies were declared in the District's manufacturing industries. Unemployment continued to rise up to September 1982 when it peaked at 16 per cent. It then remained around this level until 1985; since then the figures showed a gradual decline in registered unemployment, most notably during 1977-8.

These official unemployment statistics must, however, be treated with a considerable amount of caution. There have been three fundamental changes in the basis of the unemployment figures: in 1982 when the count changed from people registered at the government's Job Centres to people receiving benefit; in 1984 when the definition of the government areas for collecting statistics was substantially changed; and in 1986 when the government revised the figures in order to give a more accurate measure. In addition to these three major changes, there have been 16 other changes since 1979 in the method of collecting statistics. For these reasons, unemployment figures should be taken only as a very rough guide to the real unemployment situation. It is, however, reasonably safe to say that the general trend in Bradford is characterized by rapidly increasing unemployment from 1979 to 1981, followed by a levelling up to 1985, and thereafter a gradual fall.

It would be easy to conclude from the statistics presented above that Bradford's unemployment problems are not so serious as they were a few years ago. Further consideration suggests that such a conclusion would very much oversimplify a complex problem. There are, for example, major local variations within Bradford. The overall unemployment rates mask the fact that there are large areas where unemployment rates are much higher than the average. Seven electoral wards have unemployment rates about 15 per cent. Significantly, the three inner-city wards in Bradford, which together house half of the District's ethnic minorities, have unemployment rates above 20 per cent. In University Ward, for example, where two-thirds of the population are from ethnic minorities, the overall unemployment rate is 29 per cent and the unemployment rate for 16-19 year olds is 70 per cent. In small areas the differences are even more pronounced. At the time of the General Population Census in 1981 for example, the Bracken

Bank Estate, one of the District's worst unemployment black spots, had an unemployment rate of 26 per cent, while in the neighbouring village of Oakworth it was only 6 per cent.

It is essential to understand why these differences persist. What is clearly not a factor are local differences in employment opportunities. University Ward, for example, has both the highest level of unemployment and one of the largest concentrations of jobs in the District. Simply providing more jobs in local areas with high unemployment is likely to do little or nothing to tackle unemployment in these black spots. The real explanation is that the District's most disadvantaged groups have tended to converge on areas with the lowest-cost housing. The inner-city areas, with large and cheap terraced houses, have particularly attracted ethnic minorities - a group hit hard by job loss in the textile industry and often multiply disadvantaged in the competition for jobs. Council housing has traditionally provided housing for low-income groups, including unskilled workers, another group hit particularly hard in the job losses of the recession.

The category of the long-term unemployed needs particular attention. Government statistics show quite clearly that, while overall unemployment levels are falling significantly, there has been very little reduction in the number of people who have been unemployed for over a year. Yet it is this long-term unemployment which is the real issue for the Bradford District's people. For many of the long-term unemployed not eligible for unemployment benefit or any redundancy money, and with savings long since spent, unemployment means total dependency on social security and real and permanent poverty.

Many of the long-term unemployed will have been thrown out of work from the District's traditional industries in the mass redundancies during the early years of the recession and will never have worked since then. Others will have been school-leavers unable to find jobs except for the occasional temporary job or Job Creation Scheme. Because they lack any real work experience many employers take the view that they are virtually unemployable. Unemployment, and long-term unemployment in particular, affects two groups: young people and ethnic minorities. Young people are hit disproportionately hard by unemployment. Even with schemes such as the Government's Youth Training Scheme and the Community Programme making a significant contribution to keeping unemployment rates down, the rate for 16- to 19-year-olds in the District was still over 20 per cent by 1987 - double the rate for 35- to 44-year-olds. Unemployment is also particularly severe amongst the District's ethnic minorities. Although recent local figures on ethnic unemployment are unavailable, national estimates for 1985 showed unemployment among ethnic minorities at double the white rate: three times as high for people of Bangledeshi/Pakistani ethnic origin who make up about two-thirds of Bradford's ethnic minorities. For 16- to 25-year-old Bangledeshis and Pakistanis, the national unemployment rate was 48 per cent. In Bradford the comparable rate will almost certainly be over 50 per cent. This bleak picture is confirmed by the Bradford Council's Careers Service statistics which show that out of 1,047 ethnic-minority pupils who left school to look for work in 1987 only 137 - 13 per cent - actually got jobs (excluding government schemes) compared to 42 per cent for white school-leavers. Population trends mean that already 20 per cent of children reaching school-leaving age in the District are from ethnic minorities, and by 1997 this percentage is likely to rise to 30 per cent. Consequently, although there are no statistics to confirm it (the government stopped monitoring ethnic origin in relation to unemployment in 1982), it seems likely that long-term unemployment among ethnic minorities is also significantly worse than for whites.

## THE COUNCIL'S POLICY RESPONSE: THE INTEGRATED APPROACH

Bradford Council has always recognized that it could not tackle the District's problems single-handed. Hence, when the European Commission suggested that regions and cities should develop integrated strategies to tackle regeneration, Bradford seized the opportunity. The Council developed a strategy based on the need to rebuild the District's infrastructure and improve the environment, thus enabling economic growth and subsequent employment opportunities to develop. It devised an approach involving a partnership with central government and a wide-range of key public bodies such as the water and transport authorities. The aim of the integrated approach was to secure a stronger basis for economic growth, the restructuring of the local economy, and a reduction in unemployment.

In addition to high unemployment levels, Bradford

suffers from an outdated and worn-out infrastructure. There is a shortage of land, and most of the vacant sites are derelict and require substantial environmental treatment, new drainage, and new access roads. Similarly much of the vacant industrial floor space is in Victorian, multi-storey mills. The combination of industrial decline and the resulting despoiled environment has affected the District's image, which, when added to the country's North-South divide, makes it very difficult to attract private investment.

A wide range of central government measures operates in Bradford, such as regional assistance and the Urban Programme. Combined with assistance from the European Regional Development Fund and the European Social Fund, these measures led to a recent improvement in Bradford's economy. By 1987-8 the city was at a watershed, with a number of potentially large-scale developments at the design stage. It was hoped that the Integrated Development Operation would provide the impetus to a renewal of economic activity to take Bradford into the next century.

The Integrated Development Operation is a five-year rolling programme with the following strategic objectives:

to create a more favourable industrial and commercial climate for economic development.

to enhance employment opportunities and raise the standard of living of Bradford's workforce with new skills and increased productivity.

to provide a modern infrastructure and land compatible with the needs of modern industry.

to improve Bradford's internal communications network as well as links with the regional and national networks.

to maximize the positive visual impact of Bradford's dramatic and potentially attractive environment as a way of improving the area's image and boosting business confidence.

These objectives will be achieved through four action programmes which are all interrelated and which are underpinned by training schemes designed to ensure that the people who are out of work benefit from the jobs created by the construction work in the programmes and by the renewal of economic activity.

The Infrastructure Action Programme is aimed at removing development constraints from industrial sites - clearing and restoring chemically damaged land, providing improved water-supply and drainage facilities, and the basic services for power supplies. Complementing this programme is the Environmental Action Programme aimed at improving the District's image. Thus environmental work will be carried out along key road and rail corridors and in industrial, commercial, and tourist areas. The third programme is a Transport Action Programme which will improve the highways network and provide better access to industrial sites. On a different level the Transport Action Programme will also lead to better labour mobility by helping people out of work to have easier access to jobs.

Finally, the Economic Development Programme focuses on four sets of measures: providing industrial land once the infrastructure has been improved, providing modern industrial premises, assisting business creation and development, and tourism. The Land Programme is linked to the infrastructure measures, while the provision of premises will lead to an increase in managed workshop spaces and small industrial units. Assistance to businesses will focus on small and medium-sized enterprises and will provide advice and assistance through the current government measures, plus new programmes to encourage the transfer of new technology between the local university and local companies and to encourage better use of information technology.

A feature of the Integrated Development Operation is that each Action Programme has a clear set of key outputs and targets showing both the construction tasks, such as the building of 1.9 km of new roads and three new railway stations, and the employment and training areas where the targets are the creation of up to 2,000 new jobs resulting from new industrial premises and a throughput of 15,000 trainees between 1989 and 1992.

## WHY THE INTEGRATED APPROACH?

At first sight Bradford's approach appears to be the opposite to that of Strathclyde. The integrated approach seems to focus exclusively on economic regeneration and to ignore the social, whereas the reverse appears to be true in Strathclyde. In reality both local authorities have economic and social strategies. Alongside the integrated approach there is considerable work on community development in Bradford, albeit not on the same scale or depth as in Strathclyde, while work in Strathclyde on economic regeneration, especially in Glasgow, has produced a number of outstanding successes in the last ten years.

A reason for Bradford's developing the integrated approach was primarily that, as an English metropolitan district, the Council did not have the scope to effect radical changes to the infrastructure single-handedly. The advantage of a joint approach is that it would bring together all the public agencies that are responsible for the infrastructure. Consequently, it was hoped that this approach would achieve a far greater impact within a shorter timescale than the Council's carrying out worthwhile but relatively small developments by itself.

A second reason for developing the integrated approach was the realization that economic regeneration could take place without radically affecting the unemployment figures. The 'trickle-down' theory of free-market economics simply does not work in a place like Bradford. The great majority of people who are unemployed do not have the skills to obtain the new jobs resulting from economic growth. It was, therefore, essential to develop a training programme that could be integrated into the anticipated economic development, if local people were to benefit. This is especially true in areas where ethnic-minority unemployment can be three times greater than white unemployment.

Finally, Bradford's involvement with the European Community stemmed from close links which the District had developed with the Community over the last ten years and the undoubted support that European financial instruments had provided to the District in rebuilding the infrastructure.

The integrated approach was clearly focused on economic development and not community development. However, as can be seen from the first section of this chapter, Bradford Metropolitan Council recognized the problem of increasing poverty on housing estates and the reality that, even if the integrated approach were an outstanding success, there would still be people out of work and deprivation and disadvantage in many neighbourhoods throughout the District. In tackling these problems Bradford has much to learn from Strathclyde's approach to community development and in particular from Strathclyde's success in developing third-sector community businesses.

Despite the relatively narrow focus of the integrated approach, the point that Young makes in his chapter on Strathclyde 'that many staff did not know we had a strategy, let alone what it meant for them' is equally applicable to Bradford. For this reason the Council restructured its Divisions in order to create one Directorate of Employment and Environment covering infrastructure improvements, economic development, and employment initiatives; also, a training programme was developed to ensure that staff throughout this new Directorate were aware of the strategy and their role in achieving the objectives.

## CONCLUSION

Integrated operations were for Bradford a learning experience and at the same time an attempt to pursue a distinctive approach. One of the advantages of working with the European Commission was that there were a growing number of integrated operations taking place across the European Community. By adding the new ideas and lessons of this experience to other developments in Britain, such as the urban development corporations, and the lessons from the United States in cities such as Baltimore and Boston, Bradford was better placed to learn about the merits and demerits of different approaches. The most innovative aspect of Bradford's approach was the clear pathway from infrastructure improvements to training courses that would enable unemployed people to benefit from projected economic development.

In both England and the United States there have been cases of economic regeneration completely bypassing the very people who are the most disadvantaged and of public money merely enabling private companies to generate greater profits with little or no gain to the local community. During the 1980s England has witnessed a number of projects in which private companies and public agencies are working to overcome this problem. A timely reminder of this issue was recently given by David Mundy in a paper relating to the redevelopment of waterside sites. Despite the lack of derelict docks, Bradford might yet achieve economic success and reduce unemployment by following this approach. (1) A concrete example of the importance attached to this approach is the role of the Council's YTS co-ordinator as a contact officer for a multi-million project where a private company is redeveloping the city's West End.

An even greater success would be possible if Bradford could link its economic regeneration work to an integrated social strategy similar to that of Strathclyde. Whilst the District has recognized the problems of housing estates, it is unlikely that they will be radically affected by proposed economic development. The economic focus of both the central government's Urban Programme and the European Commission's initiative has led local councils to pay lip-service to deprivation and disadvantage due to lack of funds for local-authority initiatives in housing and community development.

Despite the apparently booming economy, the economic and employment problems of Bradford remain overwhelming. For the 7,379 people who had been out of work for over two years and the young people who had never had a real job, the economic boom was a cruel deception. By developing the integrated approach to the regeneration of the District's economy, and by working with a wide-range of agencies from the European Community to central government and the public utilities, Bradford Metropolitan Council hoped to renew the District and recreate in the year 2000 the former glory of Bradford in the year 1900. Whether the Council can eradicate the poverty so apparent in both 1900 and 1988 remains the crucial question.

## REFERENCE

1    D. Mundy (1988) 'Making local development benefit the yuppies', Employment Initiatives, vol. 5, no. 3, June.

# Chapter Twelve

## NATIONAL/LOCAL RELATIONSHIPS IN POLICIES TOWARDS LONG-TERM UNEMPLOYMENT

Anna Whyatt

The acceptance of long-term unemployment as a facet of European society has been a gradual process over the period since the mid-1970s. Whereas it was once felt unthinkable that unemployment should rise beyond certain levels, control of inflation has now become central to European economic thought and practice. At the same time unemployment has increased; by early 1988 it showed only a slight downturn. Long-term unemployment moved from a peripheral position in the early 1980s when it was assumed that it was a disturbing but periodic phenomenon that would eventually be picked up by the mainstream labour market or industrial policies. It came to occupy a central position, both symbolically and actually, in which all the associated problems that it exposed - skill mismatch, geographic distribution, the failure of the market economy to deliver to specific communities - could be seen for what they were. This shift was important because it helped to overcome a previous failure to recognize the interrelationship of these issues and, consequently, to recognize the necessity for integrated solutions.

Whilst these trends are showing themselves very sharply across Western Europe, they are actually of worldwide significance. In particular such trends are most sharply revealed by the urban crises. The role of towns and cities is being shifted as they reflect different effects of the new adjustments. At the same time deindustrialization is resulting in a loss of manufacturing employment and a growth in service sectors, a process which is demanding a vastly different system of industrial activity and creating massive inequalities, not just between individuals, but

227

between communities and whole towns and cities. The boom in the property field in London is one example of the way in which a massively inequitable development in one geographical area has led to consequent devaluation in other areas. Western Europe is exhibiting a shift in the world economic pattern from a contrast of rich and poor nation states to one of rich stratas of society and rich geographic areas juxtaposed with the poor and disenfranchised who increasingly are spatially located within particular areas. A recent analysis by the Child Poverty Action Group showed that some 15 million people are currently living below the poverty line in Britain, four million of whom are children. Whilst wages in the top 10 per cent of the population have risen by 40 per cent in the last five years, wages in the bottom 10 per cent have risen by only 7 per cent. And if we review the position within this context, we see that we have failed quite considerably, regardless of the achievements that have been generated in the country as a whole, to resolve these problems.

Whilst in the relatively 'prosperous' South of Britain, many inner-city areas have suffered severe decline; inner-city areas in the South-East have unemployment rates as high as anywhere in the country. London as a whole has an unemployment level only superseded by that of the North-West. Rising levels of unemployment run hand in hand with the diminution of basic services and the decline of local industry to create a ghetto of total deprivation. In many instances this process leads to a syndrome which can rapidly, over say a six-year period, lead to a total disintegration of a discrete local economy in a particular area. In the London Borough of Southwark, which is only one example of this syndrome, the Gloucester Grove Estate in the Peckham area is now a virtually no-go area where milk is no longer being delivered, where levels of unemployment are rising, and where levels of crime are rising congruent with a 30 per cent rise in unemployment. This estate, however, is only four miles from the Dulwich and Docklands area of the borough, where land values and house prices have become some of the highest in the country. Unemployment in Bermondsey, adjacent to the Docklands area, has risen to 30 per cent during the period of activity of the London Docklands Development Corporation. In the same LDDC area lowest house prices were pegged at £40,000. Yet the Docklands housing-needs survey, carried out by the London Research Centre in 1987, showed that 7.

per cent of households in Newham in the LDDC catchment area had incomes below £10,507 and 75 per cent of those in Tower Hamlets below £8,591, making it difficult to service a mortgage of over £25,000. (1)

The number of long-term unemployed in this area, an example of many like it, was 50 per cent of all unemployed in 1987. These areas often contain very high proportions of one-parent families, people on low incomes, the elderly, as well as people who are unemployed. The consequence is a detrimental effect on the local economy; the money supplies in the local areas become severely restricted with a subsequent fall-off in investment. Many recent studies have revealed the direct correlation between an increase in social stress and health problems. (2) The Scarman Report identified the lack of a co-ordinated approach in dealing with social stress which is 'deeply embodied in fundamental economic and social conditions'. (3) The Faith in the City report from the unified churches identified unemployment, together with housing, as one of the basic causes of the increase in fundamental social tensions in inner-city areas. (4) The cost of rising unemployment just in meeting social-security requirements is overtaking those costs identified as basic income requirements to keep people above the poverty line. (5) To a very large degree the concept of long-term unemployment as running like a thread through the whole of society and therefore containable is now completely false. Long-term unemployment is a very large part of a syndrome of displacement and alienation from the social and economic mainstream. This aspect of the phenomenon makes it so very difficult to displace working from a national governmental base.

The facts are not encouraging with respect to the development of a skill base, technical and managerial, and the proper level of investment that is needed to regenerate industry, both in the private and public sectors. A report in 1980 by the European Commission identified Britain as having the worst training record of any country in Western Europe, outside Ireland; and the recent report Competence and Competition shows dramatically the failure of the British economy to invest at the same level as its West European counterparts or the Japanese in the development of practical skills for its people. (6) Whilst concentrating on the development of infrastructure and property, Britain has failed to focus on the adequate development of a labour infrastructure to provide the skills and expertise that are needed in local areas.

In investment, policy has been primarily directed towards the development of the finance sector and in the regions increasingly towards retail and warehousing development and property development. Without investment in industry as a specific policy, job generation will be minimal. The resolution of the long-term unemployment problem is, therefore, not simply about changing the macro-economic climate to facilitate growth. As the stock-market collapse of 1987 demonstrated, the creation of a macro-economic climate in which reindustrialization can occur, and in which associated problems are confronted, needs to take place as a specific policy; otherwise it will not happen.

If these are the targets, it is essential to devise a policy and strategy that faces up to dealing with the fundamental problems at the local level as well as the overriding problems at the national level. For it is at the local level, particularly in the urban areas, that we see not just the failure to meet employment problems but symptoms which indicate ruptures in the social structure that may spread outside to encompass other areas as well. In Hammersmith and Fulham, an outer area of London with previously low levels of unemployment, changes in the local industrial structure have brought the level of unemployment over the period 1985-7 to 14 per cent. There is a rapid transformation taking place that will move a previously largely private-sector-dominated economy there to a public-sector-dominated-economy.

The failure to understand the complexity of this issue shows itself consistently not just in the lack of co-ordination between national and local strategies but most particularly in the nature of the relationship between infrastructural and construction policy and local employment strategies designed to resolve unemployment. Strategies developed at the national level are not resolving regional problems, nor are they likely to when they are predicated on the need to respond to very different market forces and preoccupations from those facing local communities. Strategies at the national level have achieved job gains and have achieved a diminution in the rate of inflation. They have helped those in work to be better off than they have ever been. At the same time they have failed singularly to resolve the problems of specific local economies, particularly those in the urban areas. The recent implementation of the inner city task-force strategy is a recognition that something i wrong and an attempt to start from the bottom up to meet some of those problems that are not being met by national economic policies. However, this initiative in itself is revealing the extent to which the two basic strands of policy that are being pursued start from entirely different vantage points and will inevitably conflict. What is needed is to go further, to develop a national strategy that brings together these two areas of development, recognizing that both have solutions to offer but that, without a co-ordination between them, the unemployment problem and its far reaching consequences will not be met. It is not a question of infrastructural or property-led development solutions, in the context of a national land-value framework, being more desirable than labour/industry-led policies within a local framework or vice versa. What is at stake is the failure of either on its own to resolve these problems. What is needed is a co-ordination of both policies but one that fully and courageously faces the historical failures of both - the failure of an infrastructure-led policy to create employment and stimulate industry, and the failure of a solely employment-led policy with inadequate investment to cope with the sheer scale of the problem.

The key issue is that, in seeking to achieve this co-ordination, policy must genuinely face up to the real problems of local areas: mismatch of skills and jobs, lack of training, discrimination, loss of industry, and lack of investment. It must not entertain a vain hope that other strategies with other aims - maximization of gearing on property investment, infrastructural/environmental improvement - will 'pick up' these other problems as they develop, or that a trickle-down effect will provide people with employment in the long term. Britain must develop a strategy that confronts local structural decline as part of its brief, that confronts very high levels of unemployment and discrimination in employment as part of the problem that it has to deal with, and that takes on board such problems as a fundamental decline which must be arrested rather than a marginal decline that will be dealt with as the mainstream economy picks ups.

The necessity to develop a policy that responds to the specific problems of the particular situation thus becomes crucial. This approach means an integration of labour-market, training, industrial, and infrastructural policies. Much of the European Social Fund initiatives, particularly in the regions and in those programmes directed towards special-interest groups, have been based on assumptions which basically agreed with the emphasis in this chapter. They were sensitive to local needs and allowed scope to develop local solutions. But they were often and increasingly in direct conflict with national labour-market programmes.

The second factor that will need to be considered is the development of a corporate relationship between local industry, the trade unions, and the local community and the public sector realistically to confront the economic problems faced by the local area and to evolve a unified strategy based on consensus. At the TUC Euro-conference in March 1988 Edward Heath spoke of the need for consensus at the European level - between the social partners. Success at that level is contingent on success at the local level first.

No local economic strategy to date has harmed a local economy. The evidence shows that, in all cases, the economy has been enhanced. This sort of track record cannot be applied either to 'inward investment' policies for local areas, or to development-led property strategies. (7) The cost per job of local initiatives has on average kept below an £8,000-per-job mark compared to an estimated £35,000-per-job cost for regional investment. Spin-off effects of an integrated policy are very high because the integration aims to maximize the investment potential. No qualitative evaluation has been made of the health/welfare and morale benefits in areas where integrated strategies have been developed. However, the 'hands-on' experience is that a significant and discernable improvement in local morale occurs when people see definite and tangible results of policy.

A recent OECD Report, Job Creation in the United Kingdom: A National Survey of Local Models, which looked comprehensively at the effects of different types of job-creation activity in the UK, commented:

> There is a consensus that public sector job creation (through labour market programmes) does not raise national income or the stock of stable long-term jobs ... subsidizing jobs is also regarded with suspicion. Training to improve the occupational mobility of labour is necessary. In Europe this is usually thought to be the responsibility of the state. It should be recognized that training covers a wider range of activities than conventional definitions suggest. A newer school of thought argues that it is possible to cut unemployment

in the short term at the same time as creating more favourable labour market conditions in the medium term. (8)

The significance of these initiatives for national and regional policy is also profound, since they reveal that the fundamentally competitive nature of industrial development, with one area competing against another, is essentially damaging and that for specific sectors of industry, such as clothing and textiles, if the industry is to succeed on a national basis, it must be preplanned. They also reveal at this level, as at the local level, that a confrontation between labour-market policy and industrial policy or training policy is not helpful but that an integration of all three can result in enormously beneficial programmes. They reveal the significant lack of investment capital and a massive gap between investment practice, particularly its exclusive concentration on property/land investment, which is causing increasing harm to the development of the economy and to employment prospects as a result.

Pulling the real problems into focus, however, will have significant implications for both national/local relationships and the European dimension. It will mean that some very complex issues will need to be faced. What needs to be recognized is that, whereas there has been a great deal of denigration of local solutions to unemployment, the wider development of such programmes at the national and European level has demonstrated their importance for the macro debate. If, as this chapter argues, they are the only programmes that are confronting the spin-off factors of long-term unemployment, it is important that the lessons both of the problems they reveal and the solutions they offer are closely examined. This means that such issues as the consequences of transnational planning, the consequences of the internal market, necessary changes for reindustrialization in investment policy and external investment, the respective roles of the public and private sectors, and the role of participation in management are confronted as key elements within an alternative macro-economic strategy. In a sense, the debate is a confrontation between macro-economics and implementation. Current monetarist policy accepts the necessity of unemployment as a prerequisite for growth. There are clear indications that massive levels of poverty will soon accompany unemployment as a secondary prerequisite. It is also of significance that the requirements of industry for industrial growth as opposed to financial growth are totally consistent with the requirements of a local economy for full employment; yet the two are often treated as if they were separate issues.

At present macro-economic thinking is wedded to notions of the absolute role of the market, to the necessity to operate from the top down, and to a total separation of industrial and fiscal policy. The experience at the local level reveals the fallacy of such absolutes. It is an experience that is gradually being recognized in a number of circles. The Select Committee on Employment of the House of Commons has expressed its concern at the low level of job-creating activity by the London Dockland Development Corporation. In a similar vein the National Audit Office in the autumn of 1987 expressed a view that the job-generation and training output of the LDDC was inadequate. (9) Recent initiatives promoted by the private sector in the urban areas have taken on board the necessity of providing training and business-development programmes, and there has been a considerable move in many areas of the country since 1987 to develop integrated packages of the type outlined above. A recent report by PA Management Consultants, one of the major private-sector management consultants, on employment generation argues the necessity of

> grasping opportunities and building upon the strength of local economies, and people ... (with) ... responses (which) will need to involve a combination of such actions ... going to the roots of the needs of local communities as well as local economies ... Cost effectiveness requires a co-ordinated response based on analysis of issues, appraisal of options, and practical programmes of action and their implementation. (10)

At the March 1988 conference organized by the Industrial Society in Liverpool on the inner city and its future there was clear consensus between the private- and public-sector representatives present, from all the inner-area conurbations of the UK, that the answer to urban problems lay in partnership to develop the potential of local economies confronting the actual problems of local areas. The historical dispute between the perceived relevance of macro- and micro-economic programmes is no longer ar issue clearly located within distinct political ideologies. What is now required is an imaginative and rigorous integration of macro and micro policies. Only then will policy be fully prepared to ask searching and difficult questions rather than be content simply to adjust the lives of millions of men and women to fit around conventional wisdom.

The major worry is that such realizations, if they come at all, may come too late. Britain is ironically facing a skills shortage in an increasing number of areas. London and other areas of the South-East are experiencing particular recruitment problems due to increases in house prices. Those who can take advantage of such 'opportunities' will be those who are mobile, already relatively well off, and already skilled with a possibility of transfer of skills. A large and increasing number of people will be excluded. The number of long-term unemployed is not decreasing, it is increasing. Labour-market programmes, far from helping people to get jobs, appear to take people away from the dole queue for a period, only for them to rejoin at a later date. At a seminar for local government, Tony Allen, the Chief Executive of Berkshire County Council, commented: 'If we do not manage change, change will manage us.' It looks likely, in Britain at least, and notwithstanding the growth that has been achieved, that if we do not learn the lessons of the last ten years, the results will eventually subvert all our best efforts in the mainstream economy.

## REFERENCES

1   The Greater London Council Docklands Housing Needs Survey of 1985 showed that 75 per cent of all London Borough of Southwark households had an annual gross income of £10,400. Among all households surveyed in the docklands boroughs of Newham, Southwark, and Tower Hamlets 51 per cent had no savings; only 16 per cent had savings of £1,000 or more; only 8 per cent had savings of £3,000 or more.
    In 1987 residential land on the Isle of Dogs was changing hands at £5 million per acre.
2   National Council for Voluntary Organizations (1987) Long-Term Unemployment, London.
3   The Brixton Disorders April 10-12th 1981 (1981) London: HMSO, Cmnd 8147 (The Scarman Report).

235

4 Report by the Archbishop of Canterbury's Commission on Urban Priority Areas (1985) <u>Faith In The City: A Call for Action by Church and Nation</u>, London: Church House Publishing.

5 Centre for Local Economic Strategies (1987) <u>Local Job Generation</u>, London, January.

6 NEDO (1987) <u>Competence and Competition</u>, London.

7 Docklands Consultative Committee (1987) <u>Response to the Audit Office Study of Urban Development Corporations</u>, May, claims that, far from creating jobs, developments in the Docklands have exacerbated deindustrialization, with a net loss of some 4,000 jobs. A written answer of 9 April 1987 to a Parliamentary question tabled by Nigel Spearins, MP, showed that, of 7,897 jobs created by the Docklands Development, 5,059 were transferred to the area from outside.

8 Organization for Economic Co-operation and Development (1987) <u>Job Creation in the United Kingdom: A National Survey of Local Models</u>, London: Economist Publications.

9 National Audit Office (1988) <u>Department of the Environment: Urban Development Corporation</u>, Report by the Comptroller and Auditor General.

10 PA Management Consultants (1988) <u>Economic Development in Local Authorities</u>, February.

**Chapter Thirteen**

**SOME CONCLUSIONS**

Kenneth Dyson

1    A 'state of the art' has emerged in the development of policy measures to combat long-term unemployment, notably:

individual counselling (e.g. Restart).
measures to improve motivation, especially through mutual support groups (e.g. Jobclubs).
various training courses.
temporary employment and work experience programmes (e.g. the Community Programme).
incentives and subsidies.
self-employment/enterprise development measures.

Whilst these measures are work-oriented in the sense of aiming at a return to work, they contain a social dimension (e.g. counselling and mutual support and measures directed at community development and improved quality of life).

2    The long-term unemployed are not just the problem; they are part of the solution in that their own energy and initiative need to be mobilized. Policy must accordingly promote attitudinal change by first rebuilding confidence and capacity to live in groups and then by specially adapted forms of training. Just as there are EC mechanisms to promote technology transfer, so mechanisms must be developed at the EC level to transfer experience and lessons of social innovation in combating long-term unemployment and poverty. By encouraging diffusion of experience the impediments of national and local traditions to

237

effective policy-making can be, to some extent, overcome.

3  In combating long-term unemployment and poverty local authority activities are an important source of learning experience for both central governments and the European Community. Local authorities have had to work directly with the local spatial impacts of recession and structural change in the form of concentrations of long-term unemployment and poverty. Their strategies have focused on indigenous economic development (whether to meet social needs like those of council-house tenants and environmental improvements or by making available managed workshops, loans, and business advice). Measures have included local employment initiatives, economic development (infrastructure and technology transfer) and local labour-market management. Local-authority interventions have taken such forms as attracting firms to the locality, developing tourism and the arts, providing and refurbishing industrial properties, stimulating the creation of enterprises, direct local authority employment, purchasing goods from the private sector, municipal trading, and acting as a pressure group for development aid. At the same time in many countries, like the United Kingdom, there has been a failure to integrate local initiatives into national macro-economic policy. National and local policies have tended to exist in separate spheres rather than creatively interact. On the one hand, local employment initiatives offer a low-cost and better targeted approach than other programmes; on the other, their success depends on national support being made available in the form of investment programmes. (1) Against this background it is not surprising that the actual direct impact of local authority measures has been marginal. (2)

4  Mobilizing and co-ordinating local-authority action remains a serious and deep-seated problem. The problem of long-term unemployment challenges traditional professional assumptions and working structures. Services such as housing, social work, adult education, and local economic development need to be integrated into a strategic response. Particular services may be tempted to push the issue aside as not their own; thus local economic development units tend to

deem long-term unemployment to be a 'social' problem for housing and social services to tackle. Those services that respond may do so in terms of traditional professional definitions that do not meet the problem; thus social workers tend to see the answer in individual therapy. The sheer weight of existing commitments and work loads, coupled with mounting financial constraints, makes it difficult for officers to adapt. Indeed chief officers are often remote from the problem of long-term unemployment; the phenomenon is 'lost' in the peripheral housing estates, which in turn are bypassed by the status given to inner-city initiatives. The challenge is first to recognize the problem; then to identify the need for a strategic response; and finally to see that it is necessary to work with as well as for the unemployed. Each of these steps is fraught with difficulties, not least the cultural gap between local-authority professionals and the long-term unemployed (e.g. in attitudes to education). Rhetoric can too easily outstrip reality.

5 Local authorities desperately need allies in their uphill struggle to combat long-term unemployment. Here a measure of pragmatism is required. It is important to seek out support amongst venture capitalists and the private sector; within the voluntary sector; and within the European Commission and Community. In a context within which national governments cannot be assumed to offer sympathetic support, local authorities must set their own agenda and maximize their own potential through appropriate alliances. The problems with central government are not just financial or ideological; they are also problems of negotiating with an apparatus that does not have a complementary facility for interdepartmental action. Compared with local government, central government has failed to begin to integrate its response to long-term unemployment, with the Department of Employment, the Department of Trade and Industry, the Department of Education and Science, and the Department of the Environment completely divided in their responses. Efforts at integrated programming at the local level suffer as a consequence. Hence, as Bradford's experience illustrates, hitching local action to the EC's integrated operations initiative is attractive as a way of attempting to bypass national constraints.

239

## TRAINING AND LONG-TERM UNEMPLOYMENT

A central problem at the heart of long-term unemployment is the extent of 'undereducation', 'undertraining' and 'low skill' in Western European societies. In the United Kingdom numerous studies by the National Institute of Economic and Social Research and the NEDO report Competence and Competition (3) have revealed that this problem exists at all levels. In particular, young people are trained longer, more systematically and to higher standards in West Germany and, increasingly, in France. Indeed, the training gap was widening; France was training three times as many craftsmen and two-and-a-half times as many technicians as the United Kingdom. Failures of organization and resourcing are apparent in the low availability of structured, individually adapted training. In February 1988 more than 60 per cent of Job Training Scheme trainees had never started their course properly; 49 per cent of JTS leavers found their training 'not useful'; and only 12 per cent were entered for a qualification. Of those leaving the JTS, only 7.2 per cent had completed their training. Only 17 per cent were known to have found work. Thus a key employment initiative was manifestly failing. The YTS was also not an answer to the problem of a lack of craftsmen and highly skilled workers in the new technology industries, particularly computer programmers, qualified engineers, and accountants. There has been a tendency to train for skills that are not in short supply such as retail distribution, hotels, and clerical work (even then the standards of training for clerical work to which the YTS aspired are below the accepted minimum in countries like France). Especially notable have been the failure to develop computer skills in the new JTS; the failure to develop the Threshold training scheme for programmers run by the National Computing Centre; and the funding cuts and profit-making demands on the Itecs (information technology centres).

The basic need is to move from temporary work experience to training for employment, to provide both practical skills and motivation. This shift was symbolized by the transition from special measures like the Community Programme of 1982, offering temporary work on projects of community benefit, to the Employment Training Programme of 1988; and in the renaming of the MSC as the Training Commission in 1988. The compulsions of 'workfare' and of 'pricing' the unemployed into work do not meet this need.

240

Emphasis needs to be given to the provision of high-quality training; of better advice and counselling about training opportunities; of structured, individualized training; of help to specific groups like unemployed single parents in the form of provision of child-care facilities or allowance; and of standardized records of achievement at the end of training periods. Innovative methods must be used such as getting trainees to act as if they were running a firm; intensive mock interviews on video; field visits to firms employing those earlier in training; and taking great trouble over personal difficulties, like child care. Trade unions have a crucial role in negotiating for high-quality training and not just or primarily on the payments trainees receive.

## THE CO-ORDINATION AND DECENTRALIZATION OF POLICIES FOR THE LONG-TERM UNEMPLOYED

Faced with the corporate and local crises of unemployment unleashed by economic recession and structural change, public policy took the form of a series of unco-ordinated 'leaps in the dark' at various levels (symbolized in the term 'special measures'). During the 1980s the themes of 'coherence' and 'convergence' have gained momentum. Lack of coherence has been notably apparent in the United Kingdom in the 'system' of vocational education and training; there was a failure to bring together the Technical and Vocational Education Initiative in secondary schools, vocational education in further education and the YTS. An even greater complexity of schemes existed for the adult unemployed, with over 30 such schemes in the United Kingdom. The situation was confusing for provider and unemployed alike. In the name of coherence and convergence, two initiatives were launched in the United Kingdom in 1987/8: the integration of the benefit and placement services with the creation of the Employment Service as part of the Department of Employment, pulling together under one head the Jobcentres and the unemployment benefit offices; and the Employment Training Programme, unifying over 30 schemes for the adult unemployed.

A further need is to decentralize vocational training and link it closely to the agencies and actors concerned with regional and local economic development. Not least in the interests of co-ordination, it is essential to build an

241

employment-policy dimension centrally into local and regional economic development strategy. Public policy must begin by recognizing the enormous local variations in economic prosperity and the implications for the long-term unemployed. Some areas have experienced an especially sharp local deterioration of jobs in the 1980s: for instance, Mexborough, Barnsley, and Doncaster in South Yorkshire. It is essential to focus the programmes of the Training Commission and the Department of Employment on specific local areas and regions and to work with, support, and co-operate with local authorities and other agencies concerned with economic development at those levels. This type of co-ordination of central and local government action has been better achieved in France and West Germany. There has been a notable and lamentable absence of local planning of skill retraining in the United Kingdom, particularly in the context of large corporate closures (e.g. Alfred Herbert in Coventry). The ideological conflicts of many local authorities with the MSC/Training Commission have compounded the lack of information and knowledge about local labour markets.

Far from weakening central government, decentralization can actually increase its political and material resources. (4) Central governments can benefit significantly from governing with regions with enhanced socio-economic prowess which feeds back into the total system. In this sense the existence of regions makes governing easier.

## PUTTING THE INDIVIDUAL BACK AT THE CENTRE OF PUBLIC POLICY

A consistent theme of this volume is that public policy must work <u>with</u> and for the unemployed; they are not simply an object of policy. In this respect local, community-led initiatives have been enormously important at a symbolic level, and public policy at national and EC levels must do more to encourage them (as Benington and Roseingrave emphasize). At the same time two other radical approaches would help to restore the individual to the centre of the stage: a concept of the individual as a 'consumer' of training; and a move towards a basic minimum-income guarantee.

Using the Netherlands and West Germany as models, a

strong case can be argued for a revised funding of training. The central government would issue training and vocational-education vouchers as a subsidy to training. An advantage of this system is that it avoids dependence on employers in taking the initiative to finance training. Public policy helps the individual to take the matter into his or her own hands, by providing the training incentives to the individual rather than the employer. Each individual would be allocated a grant that would be awarded to the company training him or her, provided the company matched the amount. Employees might be given a statutory right to a minimum number of days for training each year, along with joint worker-management training committees.

Even more radical is the idea of a basic minimum-income guarantee, put forward by some representatives of the New Right and the New Left and part of the programme of Michel Rocard's centre-left government in France in 1988. This policy proposal accepts and universalizes the principle of unearned income and establishes it as a right for the working poor and the unemployed (as well as for property owners and shareholders). Households whose incomes fall below a stated minimum would have their income topped up by a new form of social income, without interference with wage-setting in the market-place (as long as a large gap was maintained between the minimum income and average national income per head). Such payments would taper off as income rose; would be subject to tax; would vary with the number of dependents; and would not be conditional on unemployment, illness, retirement, or age. Several advantages flow from a basic income guarantee. It removes dependency on a system of means-tested benefits which, because of its sheer size and complexity, cannot be sensitively and humanely administered. A basic income guarantee is fundamentally a libertarian approach. It reduces the role of bureaucratic oversight and the exercise of bureaucratic discretion. Above all, it facilitates a free choice of lifestyles, including alternative modes of production and the possibility of 'opting out', and will ease the transition to a new definition of work that breaks away from fixation on the idea of paid employment as work. Public choice will also be simplified and concentrated on the hard and fundamental trade-offs between generosity of the basic income payment, the steepness of the withdrawal rate of these payments, and the cost to the taxpayer. With a basic income guarantee it would be possible to integrate social-policy judgements into economic-policy management in a more explicit manner. At present social-policy judgements tend to be distorted by the convention whereby taxation is regarded as revenue and social security as expenditure. This simplistic division clouds moral issues and judgements and makes it difficult to fuse together benefit and tax reforms (e.g. to link the impact of changes in housing benefits to changes in mortgage-interest relief). Because of the difficulties of administratively simplifying and rationalizing the existing system of social policy, more radical reform is required.

## THE EUROPEAN DIMENSION OF EMPLOYMENT POLICIES

Supply-side measures, based on deregulation and the freeing of market forces, have found strong support from the European Commission. The most notable demonstration has been provided by the political commitment, enshrined in the Single European Act of 1987, to complete the single European market by the end of 1992. This programme aims at a borderless Europe in which physical, technical, and fiscal barriers to trade will be eliminated. '1992' is designed to act as a supply-side shock to the EC economies, liberating new forces for change and involving a new climate of competition, innovation, and enterprise.

The question facing this approach is whether it will, by a trickle-down effect, act to reduce long-term unemployment and poverty. Western Europe faces the prospect, and indeed reality, of urban ghettos and a large underclass. In the interests of social balance and solidarity two developments are required: a concerted growth strategy for Europe to complement the 1992 programme; and initiatives to complete a 'social Europe'.

1   The European Commission has recognized that a concerted growth strategy is essential if the 'intolerable' unemployment rate of nearly 12 per cent is to be cut. EC growth rates of 2.6 per cent in 1986 and of just over 2 per cent in 1987 were only sufficient to stabilize this level. Even before the Single European Act individual EC countries were far more open to and dependent on each other in trade terms than the EC collectively was to either the United States or Japan. Co-ordinated fiscal expansion by Britain, France, and

West Germany was a prerequisite of breaking out of the slow-growth trap. (5) Such a transformation required institutional innovation and a major ceding of sovereignty that some national governments, notably the United Kingdom, view with intense suspicion.

2   The President of the European Commission, Jacques Delors, has stressed the importance of realizing the concept of a 'social Europe'; whilst, on taking over the presidency of the European Council in summer 1988, Greece gave full backing to this priority. Here the emphasis was on avoiding a division between insiders and outsiders, with a large minority excluded from the benefits of European integration. Already, particularly through the European Social Fund, the European Commission had been an important source of initiatives and the dissemination of 'best practice'. By 1987 the European Social Fund was devoting some 75 per cent of its heavily oversubscribed funds to the long-term unemployed, with about half that amount allocated to employment creation. The European Commission had also attempted to disseminate best practice by criticizing national labour-market policies for concentrating on finding the best candidates for job vacancies rather than helping the individuals with the greatest difficulties in finding any job at all. Its emphasis was on the need for integrated programmes offering job-seeking advice, personal interviews, places on training schemes, and job-creation subsidies. The social emphasis was also apparent in concern with the issue of reconciling employment with family responsibilities, notably the European Commission's draft directive on parental leave and leave for family reasons (blocked mainly by the UK government). The European Commission's Childcare Network had revealed the wide disparities in national policies and levels of provision, with the UK having one of the lowest levels of publicly-funded child care (six times less than in Denmark). Removing barriers to mobility of labour in a single European market means tackling differences in employment opportunities, in other words a social content to the internal market. It involves a concern with poverty as a barrier to participation in the single market.

European policy involves a delicate balancing act in

order to obtain the benefits of a flexible US-style labour market without American poverty and ghettos. A more turbulent business environment has its advantages. However, it means not just fostering a high rate of business start ups but also accepting closures. An 'enterprise culture' involves the breaking down of barriers to labour mobility and support for the job-creating role of self-employment and small firms. (6) At the same time, in line with European traditions, there is a need to assert communal values, such as social justice, against the cumulative, unintended effects of markets. One aspect is identification of a role for alternative forms of production; another is recognition of the role of democracy in developing community values. Ultimately the quality of European life depends on a balance being struck between the provision of private goods and public services in the form of amenities, schools, health, the arts, and transport. Low achievers will always need more help and encouragement than high achievers; this must be the basic theme underlying effective policies to combat long-term unemployment.

## BREAKING THE MOULD

Thinking about long-term unemployment has to move beyond the traditional confines of fiscal and monetary policies and of 'neo-corporatism' and individualistic ideas on Left and Right respectively. A 'new revisionism' is needed that combines the ideas of market dynamism with social responsibility, of enterprise with access and participation. Traditional measures of deficit spending or of fiscal consolidation and tax cutting are incapable of dealing with high levels of long-term unemployment. In particular, 'insiders' in the formal labour market, including highly paid skilled workers, must take some responsibility for the 'outsiders' (the unemployed, part-time workers, many female workers, and the handicapped). Reductions of working hours (towards the 35-hour week) could be used by trade unions to negotiate binding agreements to create new jobs in return for a willingness to accept proportionate wage cuts for higher-paid workers. Governments would in turn offer companies higher profits to tackle unemployment by substantial reductions of corporate tax on reinvested profits. In order further to safeguard unit labour costs for employers at a time of intensifying international competition, more flexible working times and use of machines would need to be introduced, particularly in the export sector. In short, the agenda of industrial relations is in need of radical reform as a part of a strategy to combat long-term unemployment. Both sides of industry must be encouraged by public policy to use productivity gains to create new jobs rather than just distribute these gains amongst themselves. By these means those in employment can begin to share work and income with the unemployed. An additional means to foster dynamism and flexibility is the development of new means of support for small businesses, including venture-capital provision. However, in order to complete a strategy for combating long-term unemployment public policy will have to cease to focus solely on the formal labour market. It must recognize and revalue unpaid but 'socially necessary' work in the family, the voluntary sector, leisure, and further education. Public policy has to embrace a new concept of work in the form of the provision of a basic minimum income. By opening up work in the formal labour market and by elevating the status of unpaid work, policy can foster the vision of a more open, flexible, experimental, and dignified society. In order to move forward in this direction some powerful vested interests will have to be challenged. Interestingly, the process has already begun within the West German Social Democratic Party. (7)

# REFERENCES

1   C. Hasluck (1987) Urban Unemployment, London: Longman.
2   J. Chandler and P. Lawless (1986) Local Authorities and the Creation of Employment, Aldershot: Gower.
3   NEDO (1987) Competence and Competition, London.
4   Y. Meny and V. Wright (eds.), (1985) Centre-Periphery Relations in Western Europe, London: Allen and Unwin.
5   Centre for European Policy Studies (1988) Two-Handed Growth Strategy for Europe, Brussels.
6   D. Birch (1986) Job Creation in America, The Free Press.
7   O. Lafontaine (1988) Die Gesellschaft der Zukunft, Hamburg: Verlag Hoffmann und Campe.

# INDEX